The Mammals of
Cheshire

The Mammals of
Cheshire

Liverpool University Press

First published 2008 by
Liverpool University Press
4 Cambridge Street
Liverpool L69 7ZU

British Library Cataloguing-in-Publication data
A British Library CIP record is available

ISBN 978-1-84631-124-6 cased

Typeset in Rotis by BBR Solutions Ltd, Chesterfield

Printed and bound by Gutenberg Press, Malta

Contents

Preface

It is one of the remarkable attributes of British natural history and ecology that so much volunteer effort can be devoted to cooperative work. Professional biologists are often better placed to carry out detailed ecological studies, usually within the restraints of three-year projects, but the combined efforts of the 'amateurs'—often in fact very professional—provide a width and longevity of study that the few professionals could never match. The complementary nature of the two sorts of study is itself rewarding.

Works documenting the distribution of animals or plants at national or local scale are one of the most obvious results of such cooperative work. It is nearly a century since the previous comprehensive summary of the mammals of Cheshire was published, but that didn't have maps. In an era of rapid change, as species spread, and as the threats of global warming promise yet greater changes in the future, mapping present distributions is absolutely essential to provide a baseline for documentation of those changes.

Coward wrote that the dormouse was very scarce and subsequently died out. This volume documents the success of its reintroduction. Harvest mice probably never died out, but their status was uncertain for a period; both natural occurrences and reintroductions are detailed here. Coward thought that whiskered bats were the most common bats in Cheshire, but the pipistrelles (and we now know there are at least two species present) are certainly more common now. Is this a change in our understanding, or a change in the relative abundance of these species?

We know more bat species than he did, and similarly have more deer than he knew. Red squirrels have gone, replaced by grey squirrels. All these and more changes are described in this volume. The range of maps succinctly summarises just how much is known about each species, but this book provides more than maps.

It presents a very readable summary of the status and biology of all the mammals present in the county, has an excellent set of drawings, and will inspire future studies. I am very impressed by what the Cheshire Mammal Group has achieved, congratulate them on the publication of this volume, apologise that many of my own recent records failed to reach them in time, and promise to do better in future. I am sure other mammalogists will also be inspired by this book, and will help to collate the records needed for a revised edition in future years that will update distribution maps and document and quantify the changes that will occur.

D.W. Yalden, President,
The Mammal Society.

Acknowledgements

As ever there is a large number of people to thank for their time and effort in the production of a work such as this. Cheshire Mammal Group is very grateful to them for their support.

- Members of Cheshire Mammal Group, for all their records and survey efforts
- Chapter and species authors: James Baggaley, Elizabeth Barratt, Sarah Bird, Cynthia Burek, Val Cooper, Eric Fletcher, Charlotte Harris, Paul Hill, Eleanor Kean, Steve McWilliam, Tony Parker, Rob Smith, Elaine Tatham, Sue Tatman and Kat Walsh
- Contributors to chapters and species accounts: Alan Bowring and Kathryn Riddington
- Paul Hill, for the maps
- Melanie Bradley, for copy-editing the book, a difficult job given the number of contributors
- Julia Mottishaw for assistance with final proofreading
- Steve McWilliam, for comments and assistance on all the texts
- rECOrd and their members
- Dave Quinn, Tom McOwat and Rob Strachan, for illustrations
- Photographers as credited

Records were obtained from countless individuals and the following organisations

- rECOrd including National Biodiversity Network data
- Cheshire Bat Group
- Cheshire Wildlife Trust
- Greater Manchester Ecology Unit
- South Lancashire Bat Group
- Individual BAP groups

with apologies for omissions.

Sponsors

Cheshire Mammal Group would like to thank the following organisations for their financial support to this project:

 Chester Zoo, the main funder, without whom this book would not have become a reality

 Bioquip/Biota

 Halton Borough Council

 Vale Royal Borough Council

 Macclesfield Borough Council

 Natural England

Introduction

Cheshire Mammal Group (CMaG) was founded in 2001 to promote the study and conservation of mammals, to raise awareness of their presence in the county and to make possible the idea of producing a county mammal atlas.

After months of canvassing by local enthusiasts, Cheshire Mammal Group held its inaugural meeting at Risley Moss, Warrington on 6 October 2001. Representatives of various conservation organisations, local government and the general public supported the meeting. From the start it was agreed that the group needed to create a formal structure in order to support funding and future projects, group promotion, mammal advice and training provision and the creation of a basic distribution atlas for the mammals of Cheshire.

A working group was created to develop the project and progress the intended publication based on similar texts available for other counties. As the project evolved, the starting point became the seminal work on the fauna and flora of Cheshire published in 1910 by the county's most eminent naturalist, Thomas Alfred Coward.

The aim of this publication has been to update our knowledge of the species present in Coward's day, and to add to that knowledge with current species, new information on life history and present distribution. In his chapter on the mammals of Cheshire, Coward recorded 46 species of mammals as occurring (or having occurred) within recent years in Cheshire and its inland and coastal waters. Even in 1910 some species such as the pine marten and dormouse were found to be rare, just as they are now. Others such as the grey squirrel were not recorded at all, despite having been released in the locality 30 years previously. For some species, the situation has hardly changed over the intervening century, while others have shown population expansion.

Some texts on mammals confine themselves to land-based species but, like Coward, we have sought to produce a comprehensive guide to the mammals present in Cheshire at the start of the twenty-first century and have included the winged forms (bats) and the marine (cetaceans and pinnipeds). Some may wonder why the coastline appears in modern-day Cheshire but, like the rECOrd biological database for the county, Cheshire Mammal Group works to the old vice-county of Cheshire (VC58). This includes the Wirral, allowing us to claim this coast as our own. Additionally CMaG covers the boroughs of Halton and Warrington, north of the Mersey.

The distribution maps that accompany the species accounts have been prepared using MapMate and details of this process are included in Chapter 3. As with any work that involves distribution maps there will be gaps and missing data sets but we felt that we had to start somewhere. The review of local Biodiversity Action Plan (BAP) targets has resulted in quantitative targets being set, so we needed a baseline from which to judge our progress over the next five to 10 years. We would obviously welcome additional records for the database, and are always interested in more historic information.

The project has been a group effort from the start. Members have trawled through their notebooks, struggled with old databases written with obsolete programmes and scoured old reports, all to gather records for the database. People have been harassed by various forms of communication to try and make them write down those interesting sightings that they carry in their heads. Members have also been busy scanning the countryside for tracks and signs, undertaking surveys and freezing to death at Parkgate, waiting for a high tide that never came, in an effort to record the elusive harvest mice reported to be present on the salt marshes, all in the name of conservation.

1. The landscape of Cheshire

Geological background

The geological diversity of Cheshire is the base upon which the biodiversity of today relies. Geodiversity is the geological setting: the rocks, fossils, minerals, the soils, landscapes and natural processes, which are operating in the area. These form the backdrop for life.

Cheshire's geological heritage is relatively simple compared with some other areas of the United Kingdom. It contains mainly sedimentary rocks and lacks the complexity produced by igneous intrusions and metamorphic rocks. Most of Cheshire is covered with a drape of Quaternary or last Ice Age sediments in the form of tills and sands and gravels. These are the youngest deposits and are unconsolidated sediments, i.e. they are not cemented.

Far older are the Triassic sandstones which give rise to Cheshire's characteristic red soils and are used in so many of its older buildings. These are sandwiched between the oldest rocks present in the far east of the county: the Millstone Grits and Coal Measures of the Carboniferous period. To the south, straddling the Shropshire border, lie the Jurassic mudstones, the youngest element of the so-called 'solid geology' made famous of course by the dinosaurs.

Scientists know the Cheshire area as the Cheshire Basin and that is exactly what it is—a bowl or saucer whose outer edges are represented by the Pennines to the east and the Clwydian Hills of Wales to the west. The Cheshire Basin connects with the East Irish

Millions of years ago	Name of period	Climatic conditions
0–2	Quaternary	Alternating glacial and interglacial
2–145.5	Cretaceous	No rock records in Cheshire
145.5–200	Jurassic	Warm temperate
200–251	Triassic	Hot and dry—desert
251–299	Permian	Hot and dry—desert
299–359	Carboniferous	Equatorial

The rock periods of Cheshire.

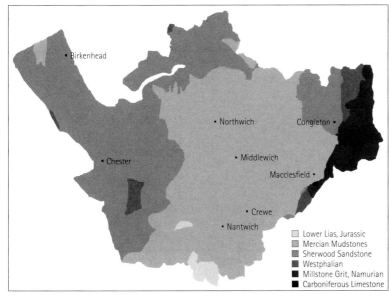

- Lower Lias, Jurassic
- Mercian Mudstones
- Sherwood Sandstone
- Westphalian
- Millstone Grit, Namurian
- Carboniferous Limestone

The solid geology of Cheshire.

Sea Basin to the north-west and the Stafford Basin to the south-east. The Red Rock Fault defines its eastern boundary, while to the west the rocks gradually lap onto the older rocks of the Welsh massif. Faults represent a vertical movement of rocks against one another generating earthquakes and the Red Rock Fault, while not very active at the present, still has the capability to move. It runs under Macclesfield and to the east of Congleton, along the Pennine scarp.

The Carboniferous period

Around 360 million years ago at the start of the Carboniferous period, Cheshire lay over the equator and most of the county was under a shallow, warm, sea full of coral reefs and fish. Land was far to the north but closer to the south. Today the coral reefs and their animals are preserved as limestone rocks in one very small area to the south of Congleton, close to Cheshire's south-eastern border. At this time, volcanoes to the south occasionally erupted ash, which was carried north on the wind and is now preserved as a tiny area of tuffs interbedded with these limestones.

At a later stage (340 million years ago) an enormous delta encroached upon this area from the landmass to the south giving us the gritstone rocks we refer to as the Namurian. As it did so the coral reefs living in the area were eventually overwhelmed and died. This time period is represented by the Millstone Grit rocks bordering on Derbyshire around Macclesfield and to the east and south of Congleton.

Moving forward in time to around 310 million years ago, the basin became land with swampy areas and flat floodplains. Trees grew everywhere and these eventually formed today's coals. This time period is called the Westphalian and the rocks are the Coal Measures. These rocks are found in four areas of

Cheshire today: in the far south-east of the county towards Stoke on Trent; hidden by glacial debris in a small area around Milton Green; in the south of the county near Malpas and on the Wirral's west coast at Neston.

The Permian and Triassic periods

The boundary between the Permian and Triassic marks the crossing from the old life forms of the Palaeozoic to the more modern life forms of the Mesozoic. As these are terrestrial sandstones there are few 'marker beds' and few body fossils preserved. This makes these rocks hard to date and the lowest beds in the area, the so-called Lower Mottled Sandstone beds, have been placed variously at the top of the Permian and the base of the Triassic. However they overlie a marked unconformity or break in the sedimentation of the area from the Carboniferous rocks to the Triassic.

The majority of the lower parts of the Triassic red beds were laid down under aeolian (windborne) conditions while later fluvial (or river) deposition and then restricted marine deposition took over. Thus the lowest beds are mainly desert dunes with enormous cross-bedded rocks exposed. The later beds contain pebble horizons as can clearly be seen in the Chester area: the so-called Chester Pebble Beds. In Chester, the Shropshire Union Canal cutting just below the city walls shows these beds beautifully, as does the picnic area in Farndon on the edge of the river Dee.

It is the well-cemented Helsby Sandstone that is largely responsible for Cheshire's backbone or the 'Mid-Cheshire Ridge', which forms the linked areas of higher ground between Helsby and Malpas, as followed by the Sandstone Trail. As we move further east, the salt deposits of the Cheshire 'wiches' (an old word meaning 'saltworks'), tell us the area was originally a

	'Group'	Names of rock formations		Conditions
		New	Old	
	Mercian Mudstone	Brooks Mill Mudstone	Upper Marls	Salt lakes and sand plains
		Wilkesley Halite Formation	Upper Keuper Saliferous Beds	Salt flats occasionally fed by the sea
		Wych/Byley Mudstones	Middle Keuper Marls	Coastal flats
		Northwich Halite Formation	Lower Keuper Saliferous Beds	Salt flats occasionally fed by the sea
		Bollin Mudstones	Lower Keuper Marls	Coastal flats
		Tarporley Siltstones	Keuper Waterstones	River, lake and intertidal area
	Sherwood Sandstone	Helsby Sandstone	Keuper Sandstone	Alternating river and desert sediments
		Wilmslow Sandstone Formation	Upper Mottled Sandstone	Mainly river sediments with minor desert sediments
		Chester Pebble Beds	Chester Pebble Beds	Large braided river
		Kinnerton Sandstone	Lower Mottled Sandstone	Mainly aeolian

Youngest → Oldest

Triassic rocks in Cheshire.

coastal plain with occasional evaporitic lakes. This is much like the Red Sea area today, where evaporation exceeds precipitation and leads to different salts being precipitated out from the hypersaline or very salty sea. Today, these are important natural resources for Cheshire and lead to unusual plant communities and habitats.

The Jurassic period

At the end of the Triassic period there was a marine transgression or invasion of the land by the so-called Rhaetic Sea. The only rocks preserved in Cheshire from this period are in the south, to the north and south of Wrenbury and around Audlem, although they are almost completely obscured by glacial till. They are represented by mudstones, signifying a quiet incursion of the sea and calm conditions.

The Cretaceous period

With the passing of time there is no evidence for what happened in Cheshire between 150 million years ago and two million years ago, as there are no rock records to help.

The Quaternary period

Geologists refer to the last two million years as the Quaternary period. This was a time of wildly fluctuating climate with long periods of cold, referred to as glacial periods or ice ages, and warmer interglacial periods. There have been as many as 20 glacial periods although there is no direct evidence for more than one or two in the Cheshire region, with the most recent being called the Devensian Glaciation. Starting over 100,000 years ago, the cooling of the climate encouraged the expansion of ice caps at the poles and in mountainous regions such as Scandinavia, the Scottish Highlands, Wales and the Lake District. The ice reached its greatest extent about 22,000 years ago after which temperatures recovered relatively quickly, and it is likely that the Cheshire region was once again glacier-free, but still very cold and frozen, by about 15,000 years ago.

The growth, movement and subsequent melting of this ice wrought dramatic changes on the Cheshire terrain. Its legacy was the landscape that our ancestors found as they returned from the south, and which they modified over the next few thousand years, first through deforestation and agriculture and, in time, various forms of industry, transport and settlement.

It is in the nature of ice ages that the latest sweep forward of an ice sheet will wipe away most traces of earlier ice sheets, so we can only guess at what effects earlier ice ages had on the region. Most likely they resulted in both erosion of the local bedrock in some areas, and deposition of material on top of it in others. In fact, most of Cheshire is thickly cloaked with till (or boulder clay). This rough mix of stones set in clay or sand has a composition that depends on the source of the material. Ice moving south from accumulation and spreading centres in Scotland and the Lake District brought with it fragments, often small, sometimes very large, of the rocks of those areas including granites and limestones. As it passed southwards over west Lancashire and the Irish Sea area, the ice also picked up sandstone and in places coal. All of these were carried in, on and within the body of ice and eventually dropped as the forward movement of the ice faltered and stopped.

Meltwater from the ice sheet found its way southward under and around the ice, carving out channels large and small into the bedrock. It also reworked some of the till and redeposited it as glaciofluvial sands and gravels.

Erosional features

The ice scoured Cheshire's soft bedrock and, in the north and west, formed several 'iceways' that aligned with the direction of flow from north-west to south-east. We now see these as the oddly shaped outer estuaries of the Dee and Mersey, the mid-Wirral trough, the Gowy lowlands and lower Weaver valley amongst others. Ice-smoothed sandstone surfaces can be seen in Wirral at Bidston Hill for example. Subglacial erosional features known as 'Nye channels' have been identified at Thurstaston and at Lymm.

In many places along the Mid-Cheshire Ridge and along the western edge of the Peak District, sinuous channels can be found which are ascribed to meltwater flow. Some are small scale, perhaps only a few metres deep and tens of metres long, whilst others are kilometres in length and many tens of metres deep. A major drainage channel cuts across the base of the Wirral peninsula, the line of which is followed by the Shropshire Union Canal. Named the 'Deva Spillway' by geologists, it is thought to have been cut by the ponded waters of the Mersey/Weaver whose exit to the Irish Sea was blocked by ice at Liverpool and which spilled across into the Dee catchment at the lowest point.

Depositional features

The larger part of the Cheshire region is covered by till of varying thickness. Interspersed with this are spreads of sand, gravel and alluvium, which are largely derived from the till. The only substantial areas free of this glacial mantle are the higher ground of Wirral, the mid- and north Cheshire ridges and the Peak fringe.

There is a number of extensive spreads of sand across Cheshire, the largest of which is the Delamere Sand. This not only underlies the forest which gives it its name, but extends east as far as Hartford, south to Little Budworth and north to Kingsley. Other similar deposits found around Congleton and in the Chelford area, for example, are extensively worked today for building sand and industry such as glass making. These sands have a variety of origins but generally represent the outwash plains from nearby ice masses. During the retreat of the ice-front during the later stages of the Devensian Glaciation, the Mid-Cheshire Ridge held ice back to the north and west whilst huge quantities of meltwater forced their way south and

east through the Delamere/Ashton gap in that ridge. A further ice-front blocked the lower Weaver valley to the north and a roughly triangular mass of well-sorted sand accumulated in the angle between the ridge and the ice-front.

Within these sands and indeed elsewhere in the county, masses of ice stagnated *in situ* as the climate warmed up from 18,000 years ago. Although there is often a reference to the ice sheet retreating, there was never any movement back north. The ice simply melted where it sat as the southward push of ice dwindled and stopped. Most spectacularly in the Delamere and Marbury areas, great holes were left in the landscape by melting ice masses. These filled with water and we know them today as 'kettle-holes', coming from an old English word 'cedl' meaning a hollow. They form dramatic landscape features and are key sites for wildlife. In some parts of Cheshire it can be difficult to distinguish between hollows formed

↓   Direction of ice flow.

╱   Extent of last ice sheet.

Maximum extent of the last ice sheet and direction of ice flow.

by this method and those caused by subsidence due to the solution of salt beds. Indeed it is likely that some have a mixed origin such as the sizeable Budworth Mere and Pickmere Hollow north of Northwich.

Many hollows, both large and small have developed accumulations of peat in the last 10,000 years. Larger areas of peat have also developed in the lowlands of the Mersey and Gowy, some of which have been exploited commercially but others remain as valuable wildlife habitats. There has also been extensive development of hill peat in the Peak fringe.

During the present interglacial, river action has eaten into the glacial mantle and redeposited these materials as relatively thin strips of river alluvium. The narrow valleys of the Weaver and Dane in particular reflect rapid postglacial down-cutting by these rivers into the unconsolidated till, sands and gravels. The rivers are characterised by steep banks cut by numerous tributary streams, and narrow floodplains.

Meanwhile, marine and estuarine alluvium has accumulated in the estuaries of the Mersey, Gowy and Dee; a process that has accelerated in historic times due in part to changes in farming practices. Also in the Mersey Valley are extensive areas of blown sand known as the thin Shirdley Hill deposits. These are located to the north and south of the Mersey floodplain, which is itself a broad band of alluvium of river/estuarine origin.

Soils

Cheshire's soils have developed over the last 15,000 years or so on the complex mix of till, glaciofluvial sands and gravels and alluvium. Together with the region's climate they have shaped the development of the natural vegetation cover that in turn has influenced Cheshire's fauna. Mankind has added further complexity and wrought further change to this mix through activities such as farming, quarrying, construction and infilling.

The future

In planning legislation, many of these sites are protected by Regionally Important Geodiversity Site (RIGS) and Site of Biological Interest status. The Cheshire region Local Geodiversity Action Plan (CrLGAP), modelled on the LBAP (see Appendix 1), aims to safeguard and maintain the geodiversity of the region and works alongside biodiversity in this respect. With climate change and raising sea levels the future for us all is uncertain.

Characteristic habitats

Meres and mosslands

Meres and mosses are an important and special habitat for Cheshire. Peat bogs form some of England's most scarce habitat and provide a unique home for a wealth of plants, animals and insects. They also provide an important feeding and stopping-off point for native and migrating birds. Because peat bogs can be thousands of years old (and predate our Stone Age ancestors) they contain layers of historical data. By examining a section of peat, scientists can tell what our landscape was like, what type of animals colonised the area, and what weather conditions prevailed.

An estimated 97 per cent of lowland bogs in England and Wales have been damaged or destroyed due to drainage for agricultural use, peat cutting for fuel in earlier times and extraction for horticultural peat. In Cheshire all areas of lowland raised bog have been disturbed to some extent, although it is thought Holcroft Moss has never been cut. Only small fragments of these original habitats remain, covering a total of 159 hectares.

Valley bogs and basin mires are distinctive features of a lowland glaciated landscape. A valley bog is distinguished by having an identifiable water flow as, for example, at Petty Pool Site of Special Scientific Interest (SSSI). A mire, however, develops where rainfall on the site is the only water supply. There is a significant group of such habitats in the Delamere complex, mostly remaining within the area of the forest managed by the Forestry Commission, including Abbots Moss SSSI. Many of the forest sites were partially drained and planted with conifers early in the twentieth century. The redevelopment of these areas as wetlands has already begun with Blakemere.

To the east of the Delamere complex there is a scattering of sites, including Wybunbury Moss, identified as a nationally important site. There is evidence that many sites were managed as local sources of peat into the nineteenth century. This habitat type is most at risk from drainage operations and nutrient enrichment.

Woodland

At the maximum extent of forest in Britain, around 7,000 years ago, most of Cheshire and Wirral would have been wooded. By 1068 less than 27 per cent of the native woodland remained. Today only a small percentage is left: woodlands are fragmented and restricted to steep slopes which are unsuitable for cultivation, and other areas which cannot be reached by grazing animals. In 1873 over 26,000 acres of woodland were present in Cheshire, and by 1980 this figure had fallen to just under 22,000 acres, of which almost a quarter is commercial forestry plantations.

In Cheshire original 'wildwoods' are represented by the woodlands in the eastern hills, dominated by oak with an understorey of rowan, ash and hazel, but with an impoverished ground flora.

In the lowlands many primary woods are found on steep slopes leading down to rivers or brooks, such as those in the valleys of the Rivers Dane, Weaver and Bollin. Dominant trees here tend to be sycamore, ash, wild cherry and wych elm.

The history of the woodlands in the south of the county is not as well documented, but they appear to have been coppice to provide fuel for the various local brine-pans. The remaining coppices in the south of the county tend to be typified by derelict hazel stools, often with oak, ash and maple. These woodlands in the Wych Valley have provided the ideal habitat for the reintroduction of the dormouse into the county.

Beech woodlands, such as those at Marbury Country Park and in the grounds of stately homes, were planted in the seventeenth century; lime, sycamore, sweet chestnut, horse chestnut—all non-native species—are often be found in such parkland settings and are home to red and roe deer at both Tatton Park and Lyme Park.

Commercial plantations in Cheshire dominate the Delamere area and provide ideal habitat for grey squirrels, wood mice and other small mammals. Older woodlands with dead and/or decaying trees provide roost sites for bats such as the whiskered bat and noctule, and are important hibernation sites for other species. Other mammals associated with our woodlands are the grey squirrel, wood mouse, badger, fox and small numbers of roe deer and muntjac. They would also have been home to the red squirrel, now extinct in the county.

In Cheshire there are two Biodiversity Action Plans (BAPs) covering woodlands: the Ancient Semi-natural Broad-leaved Woodlands BAP and the Lowland Woods, Parks and Pastures BAP.

Hedgerows and field boundaries

Hedgerows form important links between blocks of woodland (and other habitats) along which mammals can travel and feed. Large parts of the Cheshire Plain are characterised by single-species hedgerows of hawthorn, although recently planted hedgerows may also contain blackthorn, holly and hazel. Many hedgerows contain mature trees such as oak, sycamore and ash and a good mix of six or more woody species could indicate a hedge of several hundred years of age.

With the intensity of farming many hedgerows have been ripped out in some areas of Cheshire, but where dairy cattle has been the mainstay of agriculture, hedgerows have remained and been maintained. Recent agri-environment schemes such as Countryside Stewardship now provide incentives for landowners to maintain, manage and plant new hedgerows and the Hedgerow Regulations in force since 1977 also afford some level of protection.

In the eastern hills, dry-stone walls replace hedgerows, providing gaps between the stones for small mammals and predators such as weasels and stoats, although they are not confined to this habitat.

The Ancient and/or Species-rich Hedgerows and Cereal Field Margins BAPs cover this habitat, with the Dry Stone Walls BAP applying mainly in the east of the county.

Heathland and moorland

Lowland heaths are to be found in areas of the county with sandy soil, such as around Delamere, Rudheath and Thustaston on the Wirral. Grazing by rabbits is responsible for the maintenance of these habitats, but myxomatosis reduced the populations and regeneration on ungrazed heathlands resulted in scrubbing over by gorse and birch. Many of these heathlands have been planted with conifers and very little lowland heath is now left: Little Budworth Common SSSI is perhaps our best remaining example. Remnant heathland is also to be found along the Cheshire Sandstone Ridge, where the National Trust and other organisations are restoring former heathlands.

Upland heaths or moorland are restricted to the eastern hills in Cheshire where the altitude exceeds 200 m. Heavy stocking levels of sheep on moorland have reduced both the quality of our moors and also their area. The most characteristic mammal of this habitat is the mountain hare which can still be found in the hills in the north-east of the county.

The Lowland Raised Bog BAP and Heathland BAP cover these habitats in Cheshire.

Agricultural land

Most of the Cheshire Plain is under agricultural management with dairy cattle dominating the south of the county, with crops (mainly cereal and potatoes) becoming more apparent in the north. Mammals which are to be encountered in agricultural areas include the rabbit, hare, fox and badger—the latter two being more evident where good hedgerow networks exist. The house mouse and common rat are both common inhabitants of agricultural land, especially in feed stores, whilst several species of bat roost in barns and stables. Most of the grassland in Cheshire is improved, but areas of unimproved grassland and semi-improved grassland remain, some of which are home to small mammals such as the harvest mouse, wood mouse, bank and field vole. Rabbits are common here and predatory species such as the fox, weasel and stoat occur.

An important feature of the Cheshire agricultural landscape is its water-filled marl pits. Excavated to provide marl for applying to the land prior to cultivation, these ponds are to be found in most of the fields within the county—in fact Cheshire has been dubbed the pond capital of Europe!

Water voles and water shrews may both be found in ponds within the county, whilst the ponds can also act as important drinking supplies for badgers in periods of drought.

Even though agricultural land is a man-made habitat, Action Plans still exist for some elements such as the Cereal Field Margins, Ponds and Ancient and/or Species-rich Hedgerows BAPs.

Rivers and canals

As with hedgerows these provide valuable linking corridors between habitat blocks for several species, but are also habitat for the otter (now returning to Cheshire), American mink, water vole and water shrew. Initially sightings of polecats were from along river corridors where they were crossed by roads. Many of our major river systems are becoming cleaner and sightings of seals, and the occasional cetacean in the River Mersey as far as Warrington, may be a result of improved water quality and food supplies.

The estuaries of the River Mersey and River Dee, with their associated saltmarsh, provide homes for many mammal species, their numbers often only apparent at high tides when harvest mice, moles, voles and mice are pushed to higher ground, often running around the feet of observers. At the mouth of the River Mersey, the sandbars form important hauling out areas for seals, whilst fortunate observers may also see dolphins or porpoises breaching in Liverpool Bay.

There are no specific BAPs in Cheshire which cover these linear features, although sections of some river corridors will receive attention under the Reed-beds BAP and also the Species Action Plans for riparian mammals. At the coast, the Coastal and Floodplain Grazing Marsh, Coastal Saltmarsh and Coastal Sand Dune BAPs cover the estuarine ends of the county's rivers.

Reed-beds and fen

Reed-beds in Cheshire provide habitat for harvest mice, bank and short-tailed field voles, mink and occasionally otters. They are mostly small or isolated from one other, adjoining larger meres such as Rostherne and Budworth Mere, while others may be found on the Dee Estuary or along main water courses such as the River Weaver.

Rather than receiving and retaining their water and nutrients from direct precipitation, fens are peatlands fed by water coming from elsewhere. This can happen in two ways: in topogenous fens (those formed by geological features such as basins or flood plains) the water movement is generally vertical, whereas water movements are lateral in soligenous fens which are fed by ground water from neighbouring springs, rills, flushes and valley mires. Fens can further be distinguished as rich or poor according to the amount of nutrients carried in the water that feeds them.

In the Cheshire region, fens support most of our small mammals. They are often found in a complex of habitat types associated with meres and mosses as part of the hydrosere succession, such as at Hatchmere, or as part of fen woodland or fen pasture. Petty Pool SSSI contains an extensive mosaic of poor and rich fen and is considered as the best example of these communities in the county.

There are Local BAPs for both Reed-beds and Fens in Cheshire.

Urban (cities, towns and suburbs)

Gardens and parks provide vital green areas within the built environment and can be an oasis for mammals. Small mammals such as the wood mouse, bank vole, field vole and common shrew inhabit many gardens, whilst the house mouse is probably more common than reported. Hedgehogs are regularly seen and encouraged by gardeners, whilst the hills left by moles are less welcome! Badgers and foxes will often visit suburban gardens after dusk to forage and in larger gardens have been known to be found lurking under the garden shed! Attic spaces and cavity walls in modern buildings provide roost sites for colonies of common and soprano pipistrelle bats. In urban areas, the common rat is never very far away and foxes will roam the streets at night clearing up leftover take-aways or food from outside restaurants.

The Gardens and Allotments BAP is applicable to the urban, suburban and rural built environment.

Industrial landscapes

With regard to mammals in Cheshire, the industrial landscapes fall into three broad categories—the docks on the Wirral, sand quarries and the legacies of the salt industry of mid-Cheshire. The ship rat is an inhabitant of the docks, albeit a rare one, with very few recent sightings; the common rat is now more likely to be encountered there. Operational sand quarries do not afford much habitat for mammals but imaginative restoration schemes provide new woodland and grassland, all with potential for colonisation from neighbouring areas. The former deposit beds from the salt industry in the county are now mainly grassland and provide habitat for shrews, mice, voles and rabbits.

The Limebeds BAP covers the remnants of the former salt industry in Cheshire, whilst several Habitat Action Plans, such as the Heathland BAP, can be applied to restored industrial areas.

2. The history of Cheshire mammals

During the early Carboniferous period, the shallow seas that covered most of Cheshire were full of crinoids (or sea lilies), brachiopods, corals and occasional gastropods. Their body fossils (skeletons or shells) are preserved in abundance in the limestones from that period.

However, there are very few examples of body fossils from later periods. Most fossils are preserved in marine conditions but, during the Permian and Triassic periods, this area was above sea level: not only were few life forms able to survive the harsh desert conditions of that time, but any remains would not have been preserved.

Fossil evidence of Cheshire's first mammals is similarly sparse, as the county lacks the limestone caverns which can and have provided much evidence regarding the extinct mammal fauna of an area. Many of the species that may have occurred in Cheshire can therefore only be surmised from adjacent counties. The chief sources of fossil mammals in the area are from the Pleistocene epoch of the Quaternary period, with later cave deposits in North Wales, in the Cheshire and Flintshire valley gravels and in the peat-beds exposed along the Liverpool Bay coast.

Cheshire does boast good examples of trace fossils. These are not the remains of the organisms themselves, but other evidence of their existence such as trails, homes, burrows and footprints. The area's *Chirotherium* footprint trails are famous and provide evidence for the existence of dinosaurs walking across the county's flood plains. These footprint fossils can be seen in Thurstaston Country Park, Grosvenor Museum in Chester or in Liverpool Museum.

At the end of the Devensian period the British mammal fauna included the woolly mammoth (*Mammuthus primigenius*), woolly rhinoceros (*Coelodonta antiquitatis*), reindeer (*Rangifer tarandus*), giant deer (*Megaloceros giganteus*), brown bear (*Ursos arctos*) and wolf (*Canis lupis*). Most evidence for these animals comes from the caves in the Peak District and North Wales and from local peat deposits. Dating these remains is difficult and some mammoth and rhinoceros finds may belong to earlier periods, although remains from Shropshire have been dated to 12,400 years ago, some of the most recent in Britain. Remains of reindeer and bison are abundant in the Castleton area of Derbyshire; this may indicate a regular migration route from the Derwent Valley through to the plains of Lancashire and Cheshire. There are few records of giant deer although remains have been found at Wallasey.

During the period 14,000 to 11,000 years ago the climate warmed and much of the area became birch scrub. Many specimens of giant deer have been found in the Dublin area from this date, so it is likely that the animal was present in the Cheshire region at this time. This period also provides the first evidence of human

Mammoth	Adlington, Bolesworth, Coppenhall, Marbury, Mere Hall, Northwich, Sandbach, Wrenbury
Reindeer	Chester
Giant deer	Wallasey Pool

Cheshire sites with glacial period fossil remains.

Red deer	Combermere, Macclesfield, Norton, Rostherne Mere, Thornton-le-Moors, Tytherington
Roe deer	Ruston
Aurochs	Kingsley, Northwich, River Dee (Chester), Runcorn

Cheshire sites with Mesolithic subfossil remains of red deer, roe deer and aurochs.

DAVID QUINN

Wild boar.

hunting: an elk skeleton from High Furlong near Blackpool had lesions on the long bones probably caused by barbed points or axes.

The climate cooled again between 11,000 and 10,000 years ago and the giant deer appears to have become extinct during this period, although the horse, reindeer and lemming returned. The nearest remains of reindeer are from the Manifold Valley, Staffordshire and can be dated to 10,600 years ago.

The postglacial period began 10,200 years ago when evidence shows an 8°C increase in summer temperatures over a period of 50 years. This was followed by an equally rapid change in the fauna. Reindeer may have survived in upland areas for the next thousand years but red deer (*Cervus elaphus*), roe deer (*Capreolus capreolus*), elk, aurochs (*Bos primigenius*) and wild boar (*Sus scrofa*) became the main prey of Mesolithic hunters. There are no sites known locally with bones of these species but numerous undated records exist from the Mersey flood plain and may relate to this period. Skulls of aurochs and the smaller *Bos longifrons* have been found in several places including the Wallasey shore, the bed of the Mersey and during the excavation of the Manchester Ship Canal. There are few records of smaller mammals, perhaps because the habitat was unsuitable or that their bones have been overlooked. Several skulls of shrews *Sorex araneus* and *Neomys fodiens* have been found in peat-beds at Leasowe.

Submerged peat forests along the Wirral shore at Leasowe and Wallasey contain remains of recent mammals, both wild and domesticated. Most seem to be associated with Roman or Romano-British remains. However it is in the glacial alluvial deposits that the oldest remains of mammoth and a species of elephant have been found. These include a femur of woolly mammoth excavated from Wrenbury, a tooth from a sand and gravel pit at Marbury and another reported to have been found at Northwich. Leith Adams (1877–81) lists both the woolly mammoth and straight-tusked elephant (*Elephas antiquus*) as occurring in Cheshire, the latter at Coppenhall.

An increase in human interference in local ecology and the arrival of domestic sheep (*Ovis aries*) and goats (*Capra hircus*) mark the Neolithic period as new farmers moved in from the Middle East. Unfortunately, evidence of changes in local fauna is poor up to Roman times compared with southern England. Both house mice and yellow-necked mice were found in Roman Manchester, whilst remains of red deer, roe deer, wild boar and wolf were found in Roman Chester. Remains of the wolf were thought to be frequent occurrences in the Wirral peat-beds, but examination has shown the remains to be from domestic dogs. This makes it impossible to determine at what time the wolf became extinct in Cheshire forests. However, there are at least 20 place names in Cheshire that recall this species.

The wild boar existed until mediaeval times and is memorised in place names such as Wildboarclough and perhaps Bosley. Tusks of this animal have been found at many Roman sites and a skeleton is said to have been found at Mobberley many years ago.

The beaver, brown bear and aurochs had probably disappeared from the English fauna by Norman times, most likely due to hunting. It was at this time that the fallow deer (*Dama dama*) and rabbit (*Oryctolagus cuniculus*) were introduced as a source of food and fur, the former being kept in deer parks and the latter in protected warrens. Red deer and wild boar were also imparked. A total of 99 'hays' were mentioned in the Doomsday Book: these were apparently enclosures for harvesting woodland game, with roe deer being the most common species involved. The Doomsday Book also records the county town paying £45 for three timbers of marten skins, a timber being between 40 and 60 skins, presumably those of the pine marten (*Martes martes*).

Deer became increasingly important both socially and economically, and forests were designated by the king for hunting and by other nobles under licence. Most of these were established in wooded areas, but others were in treeless areas such as the High Peak on the Pennines. The remnants of this deer herd were reputedly enclosed in Lyme Park.

The Elizabethan period saw the beginning of formal control of vermin with the act of 1564 for the 'preservation of grayne'. This empowered churchwardens to pay bounties for various pest species, not only rats and mice but also many predators. This would eventually lead to the extinction of many carnivores, the pine marten being last recorded in the 1880s, the polecat in 1900 and the wildcat before 1800. The otter also suffered persecution: originally common even around the industrialised River Mersey, many were killed as the railways were laid across their wetland habitats.

One significant introduction during this period was the common rat (*Rattus norvegicus*), arriving from Russia in the late 1720s. The species rapidly displaced the ship rat and common rats are now the major vertebrate pest within the UK, whilst the ship rat has become virtually extinct.

The Victorians, as well as enthusiastically exterminating predators, also introduced species for their attractive appearance. The grey squirrel was introduced to Cheshire at Henbury in 1876 from North America, eventually displacing the native red squirrel from the county. Reeves' muntjac (*Muntiacus reevesi*) was accidentally released in 1989 whilst the American mink (*Mustela vison*) has spread through the area after releases from fur farms during the 1950s.

3. Mammal recording in Cheshire

Background

Britain, as a whole, has long been at the forefront of biological recording. This is mainly via amateur naturalists who have delighted in the outdoors and in the flora and fauna of their areas for the last several hundred years.

Prior to the Victorian period, much of this interest in natural history occurred in the gentry and clergy: these were the people with the time and money to undertake such pursuits, as well as the education to lift their interest. During the nineteenth century ordinary lay people and workers began to develop an interest in the wonders of the world surrounding them and to realise a need to 'go into the wilds' to relieve the pressures of daily life. This period saw the development of many natural history societies and specialist groups across the country, as well as the beginning of publications such as the *North Western Naturalist*. This regular journal on north-west England's natural history, produced and published by A.A. Dallman, began in 1926. Such early publications were probably responsible for kick-starting more formal biological recording. They had a wide readership and allowed naturalists to report on their findings across the area of coverage and often beyond. This led to people comparing their finds, getting together for field trips and slowly evolving towards more formalised recording of the species seen.

Mammal recording

Mammal recording, despite dealing with much larger creatures, mirrors many of the aspects of recording other groups such as insects. Some animals such as deer can be seen readily whilst out and about, whereas others, like small mammals, must be trapped in order to record their presence. Records, however, can come from many differing sources:
- Field sightings (planned)
- Field sightings (casual observations)
- Traps
- Prey remains such as owl pellets
- Gamekeeper gibbets
- Gamekeeper record books
- Pest control notes
- Flood debris
- Field signs such as latrines, burrows, feeding evidence
- Bat surveys

In many instances familiarity with the animals sought is necessary to be able to identify them. Nowadays there are numerous good photographic field guides available and many groups and societies run courses on recording methods and on mammal identification.

What to record

Recording of mammals is important for a number of reasons. We need to improve our understanding of the dynamics of populations, and how the different species interact not only with each other but with the ecological whole that makes up our environment. There is also an increasing need to standardise recorded information to enable the data to be efficiently digitised and to make it useful over a wide front. Two main types of recording need to be taken into account:
- Casual recording: where a species is sighted by accident whilst a recorder/naturalist is out and about.
- Surveying: where a species is methodically looked for at a specific site or set of sites. This can result in negative information indicating that the species was not seen during the survey. Such information can be just as important as positive sightings in understanding changes in distribution or in habitat usage.

In each of these cases there is a need to record certain elements:
- The species name in both English and Latin, if possible. This means that the species can be checked in case of transcription errors.
- The name of the site/place at which the animal was seen.
- An Ordnance Survey Grid Reference for where the animal was seen such as SJ54018210.
- The date on which the animal was seen (e.g. 27/03/2006). Separate dates for each sighting

are best rather than range dates such as 20/02/2006–30/03/2006.

- The name of the observer/recorder (e.g. Mr Steve McWilliam) with contact details if this person has not passed data through before. This allows for problems with the data to be resolved or for unusual or new species to the county to be checked.
- The name of the determiner—the person who identified the animal—if different from the observer/recorder, with contact details.

These are the mandatory elements that make up a record. Further information is always helpful to assist in defining the record, including abundance, habits (e.g. feeding, copulating) and associated species (e.g. feeding on hazelnuts).

The current data set

The current Cheshire mammal data set, held at rECOrd, reflects many different biases in recording methods and in either focus or observability. Examples of these biases are:

- The effect of focused survey work: the brown hare is a Biodiversity Action Plan (BAP) species within Cheshire and much effort has been expended in getting both the public and the farming and shooting communities involved in watching for and reporting hare sightings. This has resulted in the brown hare currently having a relatively large proportion of records. Presently the brown hare has 4.8 per cent of total records across the county.
- The effect of the cute and cuddly: the grey squirrel is looked upon by many members of the general public as being a very 'cute and cuddly' animal. Its presence in parks and gardens allows adults and children to interact with the species through watching and feeding activities. Although the grey squirrel is a relative newcomer to the British fauna it is often reported by the public and therefore represented on the data set by 8.3 per cent of total records.
- Underrecording or lack of data: some species are obviously, or apparently, underrecorded and this can be for a variety of reasons:
  1. Changes in the taxonomy: for example, the pipistrelle bat has been split into two species: one calling at 55 kHz and the other at 45 kHz. Currently there are very few records reflecting this split in the data set. Obviously this will change over time.
  2. Data not being provided by commercial concerns: a good example is the common rat which is obviously very frequent in the

county but the data set comprises only 1.8 per cent of total records. If pest control services were to provide their data this could well be in the thousands of sightings.

  3. Sensitive data: some data, particularly for protected, BAP or Countryside Rights of Way Act species is held by individuals, groups and societies and not widely shared. This means these data sets never get aggregated into Local Record Centres or distribution atlases. It is hoped that such data can be shared in the future to maximise benefits to the animals being studied, and assist with their conservation.

- Lack of data from consultancy surveys: many of the more studied mammal species in the county are surveyed because they are protected species (Schedules 5 and 8 of the Wildlife and Countryside Act). This means that many species such as the water vole, which is a BAP species and whose habitat is protected, have been heavily surveyed over the last few years. A small amount of this 'commercial' data has been made available but the majority is still held by the surveying consultancies and/or their clients.
- Exotics: there are, or have been, reports of exotic animals within the county which have usually come from escapes from private collections. The most well known of these is the feral red-necked wallaby, with the colony in the Peak District founded in 1940 by the release of five animals from a collection at Leek, Staffordshire. The early history of this population is described by Lever (1977). By 1993 the population was down to only three individuals (Harris & Yalden, 2008).
- Vagrants: many of the marine mammals are vagrants, often only passing through our waters due to mistakes of navigation. These are represented by very few records on the data set, such as one single record each for fin whale, white-beaked dolphin and blue-white dolphin.

The future

Much more work remains to be done in recording the mammals of the Cheshire region, particularly in respect of improving the 'grain' or resolution of the distribution patterns. This will, however, involve extensive fieldwork and especially the trapping of small mammals at many sites across the county. It will also be necessary to take advantage of the work undertaken by others within the county. Better relationships will have to be built to ensure that data from local authority surveys, health and safety work such as pest control services, commercial surveys, specialist societies and the like are all brought together to help develop a much more accurate picture of the changes occurring in the distribution and population levels of Cheshire's mammals. Understanding these, particularly if we are moving into a period of climatic change, may help us understand changes which could occur within our own population.

The mapping process

In order to facilitate biological recording and other scientific data gathering, the Victorian botanist Hewett Cottrell Watson devised the system of vice-counties. These are based on the traditional counties of Great Britain and Ireland, but often subdivide them into smaller, more uniform units, with exclaves considered to be part of the vice-county in which they locally lie. The use of similarly sized units provided a stable basis for recording, thereby allowing the easy comparison of data collected over long periods of time. Historical and modern data can also be more accurately compared as the vice-counties have not been affected by local government boundary reorganisations.

Watson first used the new system in the third volume of his *Cybele Britannica* (1852) to compare the distribution of botanical species across the

E	J	P	U	Z
D	I	N	T	Y
C	H	M	S	X
B	G	L	R	W
A	F	K	Q	V

Layout of tetrads within a 10 km square.

country. Although he refined the system somewhat in later volumes, vice-counties remain a standard in the vast majority of ecological surveys, even though grid-based reporting has grown in popularity.

British vice-counties are named and numbered from 1 to 112 (H1 to H40 in Ireland). For recording purposes the Cheshire Mammal Group works to vice-county 58 for Cheshire, therefore extending our coverage to the Wirral as well as to the boroughs of Halton and Warrington north of the Mersey.

The distribution maps that accompany the species accounts have been prepared using MapMate. Records were received in a variety of formats from

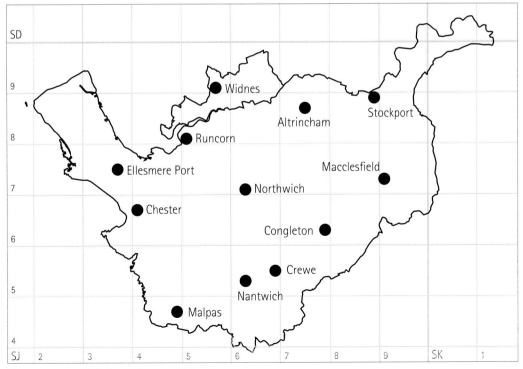

The survey area with Ordnance Survey grid lines.

observers—some as Excel spreadsheets, some as synchronisation files from other MapMate users, others from online recording systems and a small number as Word documents, emails or just handwritten notes.

The data was either entered directly into MapMate or was transferred to Excel spreadsheets ready for importing into MapMate. Data accuracy and duplication was checked, the latter using inbuilt routines in either MapMate or Excel. Any questions about the record were referred back to the recorder or the record source.

It was decided to follow the format of most other distribution atlases and produce the maps on a tetrad (2 km × 2 km) basis. Data only available on a 10 km square basis, such as that for some historical data or held by the National Biodiversity Network, was entered in the database as a 10 km square, but when mapped the record defaults to the bottom left tetrad of the square (square A).

At regular intervals maps were produced and distributed to atlas contributors in order to target areas for recording; this proved helpful in finding areas or species where records were lacking. It had been hoped that by the time of publication many of these gaps would have been filled; this has not been realised, but we do now have a starting point for future surveys.

4. Rodents Order Rodentia

Rodents form the largest order of mammals with approximately 1,500 species worldwide. Of the 17 species recorded in the United Kingdom 11 currently occur in Cheshire. A twelfth species, the coypu, has been recorded as an escape in the county.

Rodents can be identified by the single pair of incisors in the upper and lower jaw. With chisel-shaped cutting edges, the incisors are open-rooted and grow continually, resulting in the need for rodents to gnaw to maintain the size of these teeth. The lower incisors occupy extremely long sockets, sometimes extending back towards the articulation of the jaw. Rodents lack canines and there is a large gap (diastema) between the incisors and back teeth which number between three and five.

Many of our rodents have been accidentally introduced by man. These include the grey squirrel, common rat, ship rat and house mouse. These have become pest species in the UK and are often referred to (with other rodents) as vermin, due to the damage that they can do to crops and disease that they can spread.

| | Length of head and body (mm) | Tail | | Cheek teeth | |
		Length relative to head and body	Description	Number in upper/ lower jaw	Shape
Squirrels	200–300	80–90%	Bushy	5/4	Low-crowned, cusped
Dormice	60–180	80–90%	Bushy	4/4	Low-crowned, transverse ridged
Rats and mice	60–250	80–120%	Almost naked	3/3	Low-crowned, cusped
Voles	80–220	25–60%	Almost naked	3/3	High-crowned, alternating prisms

Principal characteristics of the families of rodents.

Red squirrel *Sciurus vulgaris*

Order: *Rodentia*   Family: *Sciuridae*   Genus: *Sciurus*

Status and history

The red squirrel is found throughout western Europe and has been a native of the British Isles since the last Ice Age. It is considered to have become extinct in Scotland during the eighteenth century where repopulation took place with animals from England between 1772 and 1782. However, there is a suggestion that some animals may have persisted in the north of Scotland accounting for the Scottish Highlands subspecies mentioned below.

At the start of the twentieth century the red squirrel was considered abundant in most woodlands in the British Isles, but following the release and expansion of the grey squirrel began to decline and eventually disappeared from large areas of England. Currently, populations survive in northern England, Norfolk, Cumbria, Scotland, Wales and the Isle of Wight. Most existing populations are of continental ancestry and in particular Scandinavian, which has become the most dominant form in north-east England. In Cumbria animals appear to be a British subspecies along with those in the Scottish Highlands.

In 1910, Coward describes the red squirrel as 'plentiful in wooded districts' and names various locations throughout Cheshire where this attractive

DAVID QUINN

animal could be found. How times have changed: by the mid-twentieth century the Cheshire red had become rare and localised and on the verge of extinction in the county with a few isolated records being reported from the Lymm and Rostherne area during the 1980s. They are now fully protected under the 1981 Wildlife and Countryside Act.

UK and worldwide distribution

Red squirrels are widespread in Europe, but have largely been replaced by the grey squirrel in England, Wales and in local pockets in Italy. They are absent from southern Spain and the Mediterranean islands but can be found at altitudes up to 2,000 m in the Alps and Pyrenees.

Description

Reds are better adapted to life in coniferous forests where they are able to compete with the grey squirrel, but they will also inhabit and take advantage of broad-leaved woodlands where the generous mix of seed can provide the year-round food source they need to survive. They are agile climbers and spend more time in trees than their grey cousins; they are active all year round but may stay in the nest or drey for several days at a time in severe weather. Dreys are built near the trunk of a tree or in a fork and are made up of twigs complete with leaves or pine needles, and lined with leaves, bark and moss. Animals are normally solitary but may share these nests in winter and spring and during harsh weather.

Dominance is determined by age and size and male territories are larger than those of females. When annoyed a squirrel will flag its tail from side to side and utter a harsh 'chuck', which is often accompanied by foot stamping.

Mating is preceded by lengthy energetic chases resulting in females having up to two litters a year with between one and six young in each litter. The young are born blind and deaf in February/March and again between May and August, and weaned after eight to 10 weeks. Adults can live up to seven years.

Red squirrels are primarily seed-eaters existing on a diet of hazelnuts, beechmast, acorns and conifer cones but will also eat fungal and fruiting bodies. Surplus food is randomly buried which they are able to detect by smell when needed. An animal can weigh 200–450 g depending on food availability but, in view of its smaller body size and inability to store fat, death by starvation in very cold weather does occur.

They have a range of predators including the pine marten, wild cat, owl, goshawk, stoat and fox. They are also vulnerable to the squirrel pox virus carried by grey squirrels but to which reds are not immune.

Key identification features

Red squirrels usually have russet-red fur, although coat and bushy tail colour can vary in colour from reddish brown to almost black as a result of inter-breeding with their darker European cousins. Chest and underparts are white or cream. Some reds can appear very grey and some grey squirrels can have red fur down their backs and on their feet, but the small size of the red in comparison to the stockier, rounder grey is obvious. Reds also have ear tuffs that are large in winter and gradually thin out by the following spring. There is little difference between males and females, which makes it difficult to distinguish between the sexes.

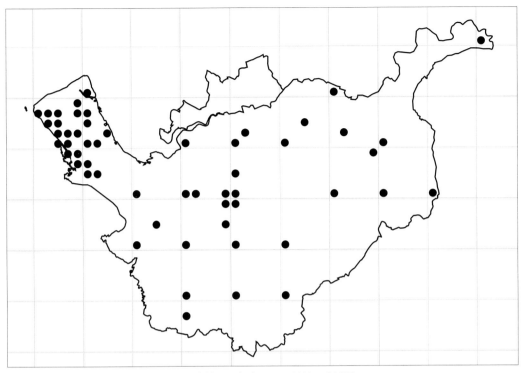

Red squirrel pre-2000. There are no records of this species between 2000 and 2007.

Observation

Where they are present, red squirrels can be observed at any time of the day especially at favoured feeding places, or close to the nest during the breeding season. It is possible to attract animals to a given place by feeding them with nuts.

Grey squirrel *Sciurus carolinensis*

Order: *Rodentia*   Family: *Sciuridae*   Genus: *Sciurus*

DAVID QUINN

Under the 1981 Wildlife and Countryside Act, it is illegal to import, release or keep grey squirrels in captivity without a licence.

UK and worldwide distribution

The grey squirrel's native range in North America extends throughout the eastern United States reaching as far north as Canada and south to the Mississippi river. Not only was it introduced to Great Britain and Ireland, but also to South Africa, Australia and Italy, where the species has extended its range into the Alps and Piedmont. It seems likely that it will now spread throughout much of Europe.

Description

Grey squirrels have a red-tinted grey fur, bushy tail and prominent ears. Adults weigh 400–600 g, head and body length is 23–30 cm, with tail length 19–25 cm. They are much heavier and more adaptable than the red squirrel with an ability to disperse widely over open countryside in order to exploit food sources and woodland habitats. They are less arboreal than the red, spending a considerable amount of time on the ground, which they cover in a series of hops and short runs often with tail held high. They have a shrill, chattering call and, when nervous, emit a scolding 'chuck chuck' accompanied by aggressive tail flicking. A territorial dominance hierarchy operates with larger, heavier animals taking precedence over smaller, younger animals. Males mark their territory by gnawing tree trunks or branches and marking with urine.

Circular in shape and slightly larger than a football, the drey consists of twigs in leaf and other coarse vegetation lined with moss, grass, leaves and even animal fur. It can be constructed at any height above ground, usually in a large fork in the tree. In summer, flimsier temporary dreys or bowers are often made on the outside branches of trees.

Mating usually takes place in December/January and again in May/June followed by a gestation period of about 44 days and lactation of up to 70 days. Females have up to two litters a year of up to seven in each litter, but usually average between two and four young. These are born naked and blind and stay in the drey until about 10 weeks. The young are sexually mature at six to 11 months. Adults have a life span of

Status and history

The grey squirrel is a native of North America. It arrived in this country in 1830 and was allowed to escape into the wild between 1876 and 1929. The first documented releases occurred in Cheshire in 1876 with further releases recorded at Woburn Abbey in 1890, Finnart in Strathclyde in 1892 and several other locations around the country by 1920. Although the released population was hit by an epidemic in 1931, it quickly recovered and underwent a major expansion of its range during the 1940s and 1950s. It is now found over much of England and Wales as well as central and southern parts of Scotland. Although still rare in Cumbria, Northumberland and parts of Scotland, it continues to expand its range.

The 1876 Cheshire release took place at Henbury Park, Macclesfield and initially seems to have gone unnoticed as Coward makes no reference to this alien species. By 1970 distribution was still patchy with animals only being recorded at Risley Moss in the north of the county as recently as 1989. Since then the species has become widespread throughout Cheshire.

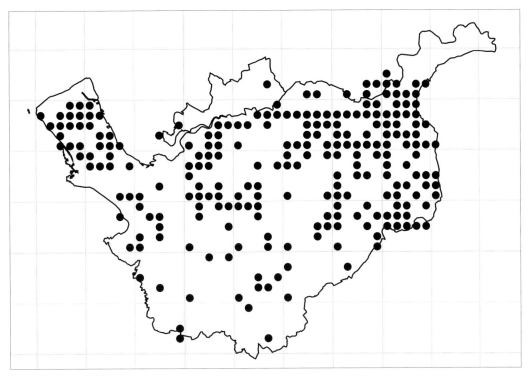

Grey squirrel pre-2000.

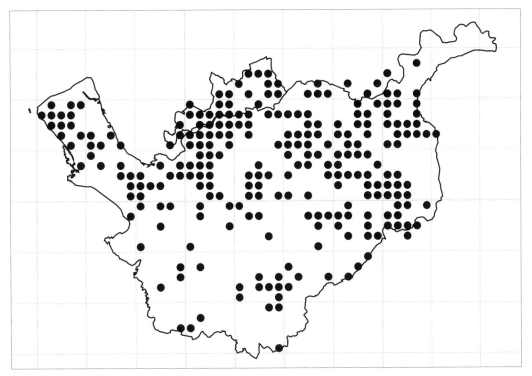

Grey squirrel post-2000.

four to nine years but dispersal of juveniles and some adults during autumn and spring can result in a high mortality rate.

The grey squirrel is highly adaptable and feeds on a more diverse food source than the red consisting of hazelnuts, buds, young shoots, tree bark, bulbs and seeds, ripe and unripe fruit and the eggs and young of songbirds. They also have an enhanced ability to digest acorns—a major food source in their native American woodlands. They randomly store seed when it is abundant and relocate these hoards by smell. Being much stockier than the red, they are also better able to survive cold winters.

Predation in the UK is minimal, but owls and various raptors can take them, with young or sickly animals falling prey to stoats, foxes or cats. They also carry and transmit the squirrel pox virus, to which they are immune but which is lethal to the red squirrel. The grey squirrel can be a major forest pest due to its habit of debarking trees and its appetite for emerging bulbs and shoots in woodland. They can also have an important impact on open nest birds, often killing the host and young or taking their eggs, as well as competing for important food resources. Population control is undertaken on fairly large local scale but impact has a limited, short-term effect on numbers and tends to occur in areas adjacent to red squirrel colonies.

Key identification features

Grey squirrels are distinguished from red squirrels by their grey fur, smaller ear tufts and their larger, more robust build. Winter fur is dense and silvery grey with a brown tinge along the middle of the back; summer fur is yellowish-brown. White underparts remain throughout the year. Grey squirrels sit with their large bushy tail arched over the back.

Observation

Common in deciduous and mixed woodland, they are also found in hedgerows, trees, parks and gardens. Bird feeders are especially liked and this animal is a regular visitor to feeding stations set up in parks and nature reserves.

Hazel dormouse *Muscardinus avellanarius*

Order: *Rodentia*   Family: *Gliridae*   Genus: *Muscardinus*

DAVID QUINN

Status and history

Hazel dormice disappeared from Cheshire at the beginning of the twentieth century; Coward notes the species as scarce. The last recorded sighting is from 1910, at Wistaston near Nantwich, although there is a report of this species in the county from 1957. They returned in 1996 when a population was reintroduced at a patch of ancient woodland in the Wych Valley on the south Cheshire border. This was the third reintroduction in Britain, previous releases having taken place in Nottinghamshire and Cambridgeshire. Over two years 54 captive-bred animals were released in Cheshire; population monitoring showed that the dormice thrived, gradually increasing in numbers and area occupied. It is difficult to estimate the total population size, but by the end of 2007 it is likely that more than a hundred animals entered hibernation.

Historically, the biggest threat to dormice has always been loss and degradation of their woodland habitat. Dormice need large areas to survive and there are few broad-leaved woodlands of the necessary size left in Cheshire. Fragmentation of the remaining woodlands is also a problem. Dormice are poor colonisers as they are reluctant to come down to ground level (except to hibernate) and will not cross open spaces. Although they will travel along hedges, hedgerow loss plus poor management further reduces their ability to move between woodlands.

Hazel dormice are offered protection under the Wildlife and Countryside Act 1981.

UK and worldwide distribution

Hazel dormice are found across the whole of Europe except the northern half of Scandinavia. On mainland Europe there are several other dormouse species, but the hazel dormouse is the only one native to Britain (the edible dormouse, *Glis glis*, has been introduced but is confined to a small area centred in the Chilterns). There are a still a few isolated populations in north-east England, Yorkshire and Cumbria, but most dormice are located south of a line from the Mersey to the Wash. Even in the south of the country, populations are often widely scattered, and the dormouse is still considered at risk.

Description

The dormouse is a woodland specialist and an agile climber found in deciduous woodland, hedgerows and scrub. The habitat must have a wide diversity of tree and shrub species to provide a range of food sources through the year. Woodlands with an open structure, such as overgrown coppice, are ideal since light getting down to the shrub layer encourages abundant fruiting of dormouse food plants such as hazel and bramble. High forest with a closed canopy can support dormice, but at lower densities than coppiced woodland.

At one time it was thought only ancient woodland provided the right conditions for dormice but they have more recently been found in secondary woodland and even in conifer plantations. It is still not clear how they feed in this environment, but dense conifers may offer better shelter from the extremes of weather than broad-leaved woodland. Conifers can also support a high volume of invertebrates, which the dormice may be using as a food source. Hedgerows are important as dormice can use them as corridors to move between woodlands, but a hedgerow is unlikely to be able to support a permanent dormouse population unless it is at least 4 m high and wide.

Hazel dormice eat a wide variety of foods, including flowers, fruit, seeds, nuts and invertebrates. Many of their food sources are ephemeral, lasting only a few weeks each year, so they have to switch frequently between food types and locations. As they emerge from hibernation in spring, hazel catkins are one of the earliest food sources. Later they feed on the flowers of blackthorn, hawthorn, bramble and honeysuckle, eating the pollen and nectar. Over the summer, invertebrates such as caterpillars and aphids probably make up much of the diet and in the autumn blackberries, wild cherries, rowan and hawthorn berries and hazelnuts are all eaten. Nuts in particular are valuable as the high fat and protein content helps dormice fatten up before hibernation, but it is believed that they avoid acorns due to the high tannin content.

Dormice are nocturnal; their big eyes and long whiskers equip them for finding their way in the dark. They are good climbers, capable of reaching the slenderest twigs and leaping between branches. Their activity is partly limited by the weather as their fur is not water repellent and easily gets waterlogged, so dormice often stay in their nests on wet nights.

They are solitary during the breeding season, but in the spring and autumn it is not unusual to find two or more dormice sharing a nest. Each animal has a home range a few hundred metres across. Males are believed to be territorial and defend their home ranges. Dormice occur at low densities, rarely more than 10 adults per hectare; this is probably because each animal needs a large area to guarantee its food supply.

Nests are built to rest and breed in; the nest being a compact woven ball of fibrous material, often strips of honeysuckle bark, or sometimes dried grasses. There may be leaves on the outside, possibly for camouflage purposes and they may be at any height, hidden in hollow tree cavities or dense tangles of ivy or honeysuckle. Dormice will readily use artificial nest-boxes and in habitats with few natural nest sites, such as young coppice, providing nest-boxes may increase the population density. Obviously nest-boxes also make it very easy to locate dormice and monitor populations as they appear to be relatively unstressed by human disturbance.

Dormice are most famous for sleeping! Hibernation is a strategy for surviving through the winter months when food is unavailable. It is far more profound than sleep: the metabolism slows right down and body temperature drops to a couple of degrees above freezing. In this state the animal needs very little energy and can survive on fat stores laid down in the autumn. If sufficient food is available dormice can accumulate body fat very quickly before hibernation, gaining up to 1 g per day to double their summer weight.

The timing of hibernation depends on food availability and the weather. Adults hibernate first as soon as they have acquired enough body fat but it may take juveniles several weeks more to gain enough weight. Hibernation can start between October and December and may be triggered by frosty weather. Emergence from hibernation in spring also depends on weather conditions, especially temperature, and can be any time from February through to April.

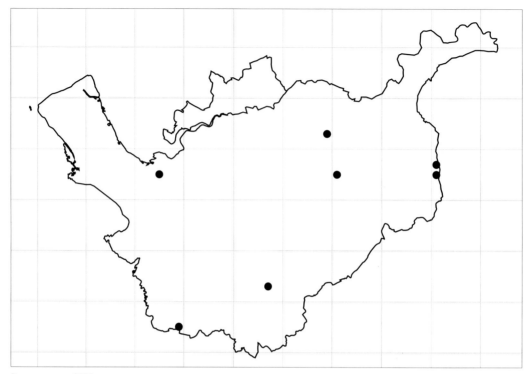

Dormouse pre-2000.

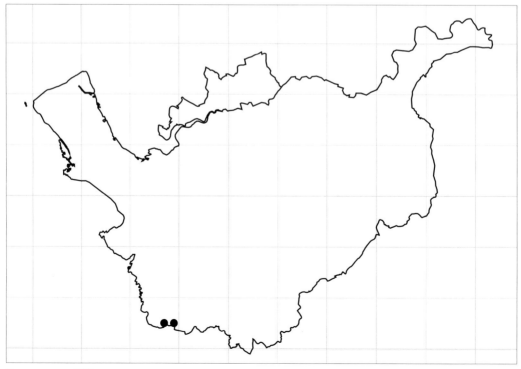

Dormouse post-2000.

Hibernation occurs at or below ground level and this is thought to be due to less variation in air temperature and humidity, so they are less likely to rouse too early in the year. They build a hibernation nest, similar to those used in the summer, which can be down a disused mouse hole, in a crevice, under a pile of logs or brash, or even just under dead leaves. While other species of dormouse (such as the edible dormouse) are known to hibernate communally, it is believed that hazel dormice hibernate alone, but little is known about this behaviour and hibernation nests are rarely found. Hibernation is a risky business for dormice: it is estimated more animals die during hibernation than from causes such as predation and disease.

A dormouse can go into torpor for several days, or just spend part of the day in torpor and then be active as normal at night. A dormouse in torpor feels cold to the touch and will sometimes breathe noisily, taking up to 20 minutes to fully rouse, even when disturbed by humans.

Dormice are unusual among small rodents in adopting a strategy of having relatively few offspring, but investing much care in each youngster. Each female has one, or at most two, litters each year, and the average litter size is four or five young. Breeding activity starts surprisingly late in the summer. The young are born blind, pink and hairless; within a week they gain a fine coat of grey hair. They grow rapidly, but will stay with their mother after weaning until they are eight weeks old. It is believed that as they grow older they accompany their mother on foraging trips. The breeding season continues into September or later and many females will have a second litter at this point. This is something of a gamble as youngsters born late in the season must grow and fatten up sufficiently before hibernation. If the first frosts come early they are unlikely to survive.

Although dormice produce fewer young than other rodents such as wood mice, many more of them survive. To further compensate, they also live longer, up to five years in the wild.

Key identification features

The hazel dormouse is easily recognised by its ginger-brown fur, fading to white underneath. It has a short face, long whiskers, small ears and large black eyes with a head and body length of 60–90 mm. The tail is bushy and nearly as long as the body with a white tip on some animals. Weight varies through the year: during the summer adults weigh 15–20 g, but in the autumn with body fat gained prior to hibernation this can double. The heaviest dormouse recorded in Cheshire weighed 44 g.

Observation

Dormice are very hard to detect in the field as they are nocturnal, arboreal and occur at low densities. They do not forage at ground level so are never caught in conventional Longworth traps unless these are set up in the branches. Predation is not a major source of mortality and dormouse remains are rarely found in owl pellets, but in places where houses are close to woodland, dormice have been recorded as prey brought home by domestic cats.

The other important field sign for dormice is chewed hazelnuts, as they leave very distinctive tooth marks. The opening is round with a smooth inner edge and diagonal tooth marks around the outer edge of the hole. The best time to find dormouse-chewed nuts is September and October, before the leaves fall and when the nuts are still pale in appearance; by spring the nuts are often decayed so the tooth marks are less clear. Due to the small numbers of dormice and the small number of nuts they eat, it can take much patient searching to find one chewed by a dormouse, even when hazel is abundant.

Bank vole *Myodes glareolus*

Order: *Rodentia*   Family: *Cricetidae*   Genus: *Myodes*

DAVID QUINN

Status and history

The bank vole is widespread throughout much of Europe and has been a resident in mainland Britain since the last Ice Age. Coward considered that the bank vole was not recognised in the county until 1889, when several animals were trapped near Northenden for the first time. The species is currently common and widespread in Cheshire.

UK and worldwide distribution

The bank vole can be found up to a height of 600 m throughout much of England and Wales, and the lowlands of Scotland including many of its islands. It is also found in southern Ireland where it was accidentally introduced in the 1950s. The species ranges across central Europe and east into central Asia.

Description

As the smallest of the vole species in Britain, the bank vole is probably one of our most common small rodents. It is found in thick cover in mixed deciduous woodlands, scrub and hedgerows but also in conifer stands and plantations where it makes its runways and burrows. Animals will also move into grassland where the population of field voles is low or absent. Social standing is determined by size with dominant males having larger ranges than females and smaller animals. Adult males are very aggressive during the breeding season and can inflict serious wounds on opponents. Bank voles have good hearing, sense of smell and homing ability. Both sexes have well-marked territories where they are active throughout the day and night, with peaks of activity occurring at dusk and dawn. In winter, activity is greatest during daylight hours. They make runways above ground in low cover but will also dig burrows, which they may share with subordinate animals.

DAVID QUINN

The bank (left) and field vole compared.

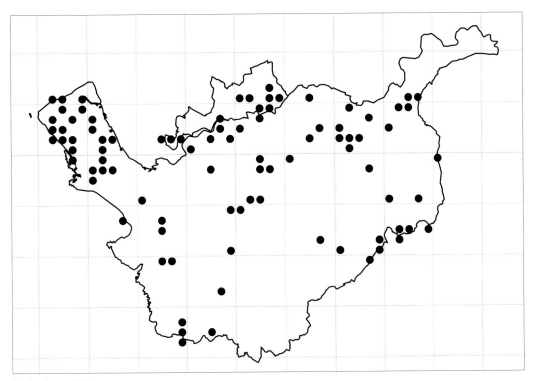

Bank vole pre-2000.

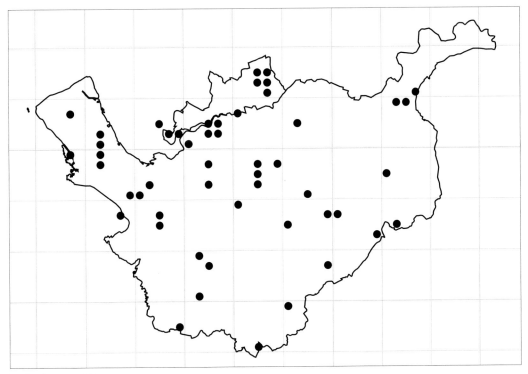

Bank vole post-2000.

Bank voles breed between March and October and occasionally in winter if conditions are mild. Females build a nest of leaves, grass, feathers and moss in the ground or in a tree or wall and, following a gestation period of 18–24 days, can produce a litter of three to six blind and naked young. They are capable of producing successive litters at three to four week intervals and conceive again whilst still suckling the previous litter. Population size can fluctuate dramatically, with late summer animals consisting almost entirely of animals born that year. Animals that overwinter produce their first young the following spring.

The species is omnivorous, feeding on a wide variety of seeds, fruits, fungi, moss, roots, worms and insect larvae and they will climb into shrubs and small trees to seek these out. They can be an occasional forest and garden pest because of their habit of debarking trees and eating seeds and seedlings. In winter, the species shows a strong preference for dead leaves, preferring those of woody plants to herbs. Excess food is stored beneath the litter layer or in burrows. In good 'masting' years populations can increase quickly over a two- to three-year period, resulting in overcrowding, stress and sudden population crashes. Bank voles have many predators including the kestrel, fox and weasel, but they are a particularly important food for the tawny owl and can form over 20 per cent of its diet.

Key identification features

The bank vole is distinguished from other voles by its chestnut-red upper parts, creamy underparts, prominent ears, rounded nose, long tail and stocky body. Adults have a body length of 80–120 mm and can weigh up to 40 g with a life span of up to 18 months.

Observation

Bank voles can be seen during the day if the observer is patient. Feeding signs or identification of remains in owl pellets best indicates their presence.

Field vole *Microtus agrestis*

Order: *Rodentia*   Family: *Cricetidae*   Genus: *Microtus*

DAVID QUINN

Status and history

The field vole is widespread throughout Europe and the only member of the *Microtus* family found on the United Kingdom mainland. Although absent from parts of the British Isles due to separation of the mainland in the Pleistocene period, it is still considered to be the most common British mammal.

In Cheshire the species is widespread in general and very common in some localities; a situation little changed from Coward's description in 1910. The population is subject to considerable variation at a localised level and this can have a significant impact on various predators as well as allowing bank voles to move into areas normally frequented by the field vole.

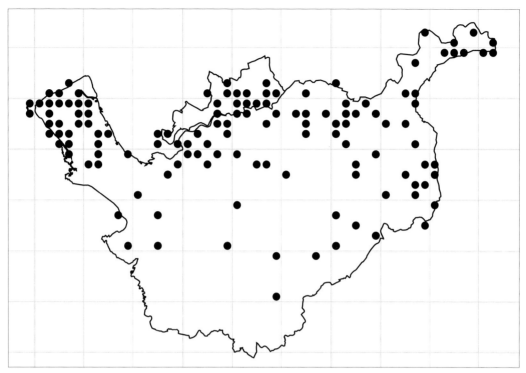

Field vole pre-2000.

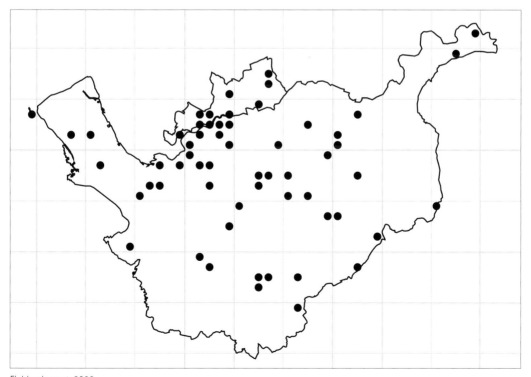

Field vole post-2000.

UK and worldwide distribution

Field voles are present on mainland Britain, but are absent from Ireland and some islands. They are found across Europe and also east into Asia.

Description

Field voles are found in rough, ungrazed grassland and sometimes in scrub and dense cover up to a height of 1,000 m. Their presence is characterised by a network of tunnels made on or below the surface of the ground, and the cut grass stems and green droppings found at regular intervals along these runs. Most activity occurs at night but field voles will also emerge during the day for short periods. Larger, stronger and older animals become dominant over weaker younger animals with males defending exclusive territories whilst those of females tend to overlap. Although several individuals will inhabit an area, the tunnel system in the litter zone is the territory of the dominant individual.

Nests are made of frayed grass leaves and built in excavated burrows or at ground level beneath a thick mat of grass. The young are born blind and naked between March and October but also in winter if conditions are favourable. Females will have a succession of litters producing between four and six young on each occasion. Lactation takes 14–28 days with those young born earlier in the year able to produce litters in the same year. Animals rarely survive for more than one year, but high fecundity levels can lead to rapid increases in population numbers.

Field voles are highly specialised herbivores feeding mainly on green leaves and stems of grass but they will also consume roots, bulbs, bark and fungi when food is in short supply. When food is plentiful, populations can increase significantly over a three- to four-year period leading to high levels of stress, poor breeding success and high mortality rates. This eventually leads to a population crash typical of vole species in general and this species in particular. The cyclic effect of these increases and declines can have a serious impact on the breeding success on some of its main predators such as barn owls, tawny owls and kestrels. Other predators include the fox, weasel and cat, which can between them take over 80 per cent of the population.

High densities can also have an impact on the amount of damage caused to grasslands, young plantations and cereal crops, but this is not usually a problem unless the population reaches a certain level. This has not happened in this country in recent times.

Key identification features

The species is recognised by its greyish brown fur with pale grey underparts. They have small eyes and ears, a blunt nose and short tail. Head and body length is 100 mm with a tail 40 mm long, and adults can weigh up to 40 g.

Observation

Often the existence of runs through vegetation is the only indication of the presence of this species. They may, however, congregate under pieces of corrugated iron left amongst the vegetation.

Water vole *Arvicola terrestris*

Order: *Rodentia*   Family: *Cricetidae*   Genus: *Arvicola*

DAVID QUINN

Status and history

Coward described the water vole as being abundant and noted records at Pickmere and the banks and marshes of the River Dee. He commented that the species was often seen around the meres and marl pits of the county. The up-to-date distribution map confirms these observations, with records widely spread throughout most of the county. However, this may mask the true status of the species, which has suffered a long-term national decline. Preliminary work undertaken in parts of the county by the Water Vole BAP Group suggests that water voles have disappeared from a number of previously recorded sites in Cheshire.

UK and worldwide distribution

The species is widespread throughout Europe, from the Mediterranean mountains to the Arctic coast, east through Siberia almost to the Pacific, south to Israel, Iran, Lake Baikal and the mountains of north-west China. Water voles are not found in central and southern Spain, western France and south-west Italy.

In the British Isles they are absent from Ireland, most islands (except Anglesey, Bute and the Isle of Wight) and much of Scotland, South Wales and south-west and north-west England. The population in north-west Scotland is considered to be local in distribution. Upland areas may be inhabited up to 1,000 m. Based on data generated from both national and local surveys from the Wildlife Trust, Environment Agency, Vincent Wildlife Trust, English Nature, Countryside Council for Wales, Scottish Natural Heritage and the National Biodiversity Network, the 2005 National Distribution Map shows strongholds in Suffolk, Norfolk, Hampshire, Essex, Caernarvonshire, Anglesey, South Lancashire and Derbyshire. All of these areas appear to be linked, at least on the 10 km square scale, by records in adjacent vice-counties.

Description

Water voles occur mainly on well-vegetated lowland ponds, slow-flowing rivers, canals and drainage ditches. They particularly favour those with water deeper than 1 m and steeply sided banks suitable for burrowing. Field signs are usually recorded within 5 m of the water's edge. The species is active during both the day and night, but evening can often be a rewarding time for water vole spotting particularly at sites with frequent disturbance.

Water voles are almost exclusively herbivores with grass forming a major part of the diet; most types of vegetation are taken, with 227 plant species being identified from feeding remains. Bark, rhizomes, bulbs and roots are also eaten and animal food such as fish, snails and crayfish are occasionally taken. Water voles consume 80 per cent of their body weight each day.

Outside of the UK, the species is less strongly associated with aquatic habitats and frequently burrows like a mole in pastures. This fossorial or burrowing form is significantly smaller in size than UK animals. The burrowing habit has occasionally been recorded in Lincolnshire, Read Island, Humberside and some of the islands in the Sound of Jura, north-west Scotland. Nesting occasionally occurs above ground in reed tussocks. Spherical nests made by water voles from gnawed reeds and flag iris are noted by Coward as being recorded at Pickmere in 1887, and at a pond in Croughton near Chester in 1907. Females, their daughters and unrelated males may nest together during the winter. Males usually leave their parents' territory at four months of age and disperse up to 230 m.

Water voles swim and dive readily, being able to trap a layer of air in the coat for insulation when wet and having skin flaps within the ear to prevent water entering. However, the species does not show a great deal of adaptation to the aquatic environment: water voles are not particularly strong swimmers, lacking webbed feet and the ability to use their tail for swimming, and their fur can rapidly become waterlogged.

Female water voles are territorial, occupying linear territories of 30-150 m depending upon habitat quality, season and population density. Many more animals can be supported per length of watercourse when habitat quality is high. Males are less territorial and their home range can overlap that of several females, with males competing for access to females. During the breeding season the edges of territories are marked by latrines and scent-marked by the male in order to avoid unnecessary aggressive encounters.

Moulting takes place twice a year with the main moult being in the autumn; individuals surviving the winter then moult again in spring. The few animals that are able to survive a second winter do not moult again and the coat becomes grey and thin. Water voles do not hibernate, but do become less active above ground during the winter, spending their time feeding on cut vegetation stored underground.

Predators include the fox, mink, domestic cat, otter, stoat and grey heron, with the mustelids being the most important. Extensive reed-beds have been shown to offer water voles some protection from predators.

Sexual maturity is usually reached after the first winter, with breeding taking place between March and October. High winter mortality means that populations are at their lowest during January/February followed by a peak in September. Gestation takes 20 to 22 days, with young born naked and with their eyes shut. Successive litters of between four and six offspring are born only six weeks apart, but 65 per cent mortality prior to weaning means many individuals are very short lived indeed. Offspring first emerge at 14 days old and both parents are involved in the care of the young.

Maximum life span in captivity has been recorded at five years, but averages only 5.4 months in the wild. Winter mortality may be as high as 70 per cent and juveniles need to attain a mass of 170 g in order to survive.

Key identification features

The water vole is the largest British vole, being larger than the field or bank vole by some considerable margin. An adult water vole has a mass of 225–310 g making it roughly rat sized, with the male being slightly larger than the female.

Due to similarities in size and common name the water vole is sometimes confused with the common rat, but the two species can be easily distinguished. The water vole has a blunt muzzle, smaller ears partly hidden in the fur and a shorter furry tail, whilst the common rat has a more pointed face, large prominent ears and a long naked tail. Guard hairs are long in the male water vole, but shorter and denser on the female.

The species usually has a brown shaggy coat with some red showing through on the back, however variations in colour are common. Melanistic individuals (having increased black pigment), are particularly common in Scotland where they can amount to 97 per cent of the population; complete albinism (absence of pigmentation) is rare, whilst partial albinism in the form of white patches on the forehead and chest is quite common. Coward noted melanistic specimens recorded in the area of the Dee in the late 1890s and early 1900s, as well as a true albino specimen observed at Micker Brook in July 1901.

Observation

Whilst quiet vigils by the waterside will be rewarded with clear sighting, water voles are more usually recorded by their field signs. These include:

- Burrows: 4–8 cm in diameter, usually in steep-sided banks near to the water level, associated with a second vertical burrow on the bank top.
- Grazed lawns: small areas of intensively grazed vegetation, occasionally found around burrow entrances.
- Feeding remains: pieces of vegetation are often carried to favoured feeding spots where remains accumulate in neat piles. Look for stems 10 cm in length often cut at a 45° angle.
- Latrines: piles of blunt-ended odour-free droppings, 8–12 mm long by 5 mm wide, deposited to act as territorial markers. Look for latrines near burrow entrances and on flat rocks and platforms of bare ground at the water's edge.
- Footprints: easily confused with those of the common rat. A clear footprint in a good substrate is needed and other field signs should usually be sought for confirmation.

Lindow Common Local Nature Reserve, off Racecourse Road, Wilmslow is a good place to look for the species.

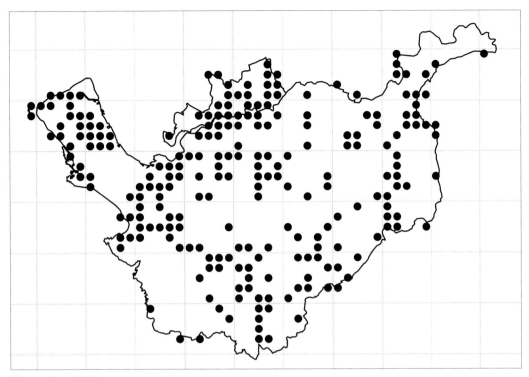

Water vole pre-2000.

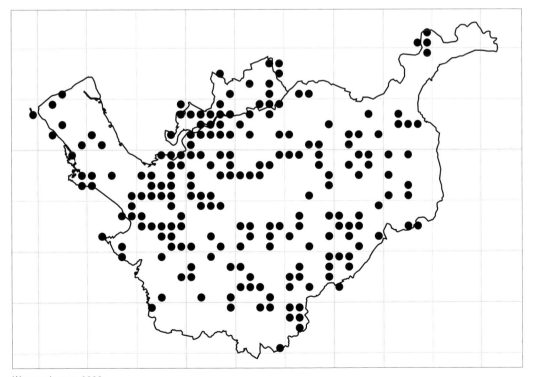

Water vole post-2000.

Wood mouse *Apodemus sylvaticus*

Order: *Rodentia*   Family: *Muridae*   Genus: *Apodemus*

DAVID QUINN

Status and history

Coward describes the distribution of this species as abundant throughout the county, and comments on how easy the mouse is to trap. Wood mice have been recorded from all but three of the 10 km squares in the Cheshire vice-county, but there is no reason to suspect that they do not occur in every square of the survey area.

UK and worldwide distribution

In the UK it is a ubiquitous species recorded through England, Scotland, Wales and Ireland, normally below the tree line. It is present in Europe within the woodland steppe zones (but not extending into the coniferous northern zone) and extends east into Asia in the Himalayas and south into northern Arabia and North Africa. It is also present in Iceland.

Description

Although they are characteristic of woodland, living amongst the ground litter, wood mice occupy a wide variety of habitats provided there is some ground cover. There appears to be competition with voles for habitat occupation: where bank voles are present, wood mice tend to avoid bracken and bramble areas and similarly their population is less where field voles are present in grassy and hedgerow-type habitats.

The breeding season for wood mice extends from April to the end of August, during which time up to four litters of up to five or six young are born. Gestation takes 25 to 26 days and the young are weaned at about 16 days old, just two days after the eyes have opened. Sex ratios at birth are approximately equal but in the winter males outnumber females. If surviving the first winter, wood mice can live for 20 months in the wild.

The nest is usually built underground and constructed of leaves or shredded grass. A system of underground runways is used to connect food stores. Food is predominantly seedlings, buds, nuts and arthropods.

Wood mice are very agile and can leap quite high, easily jumping out of buckets or tanks in which they may be placed during trapping sessions.

The wood mouse features as prey for several predatory mammals, smaller birds of prey and owls. They can make up 30 per cent of the prey taken by tawny owls in deciduous woodland.

Key identification features

Wood mice are best recognised by their large ears, black beady eyes and long tail. In adults the upper parts are

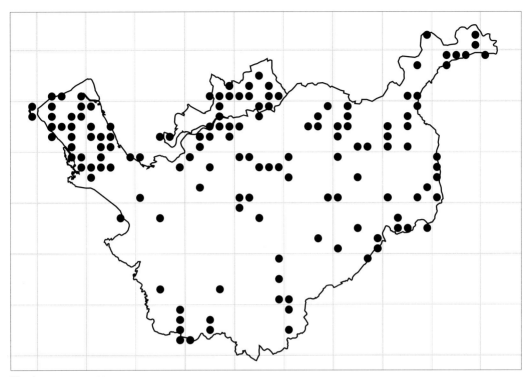

Wood mouse pre-2000.

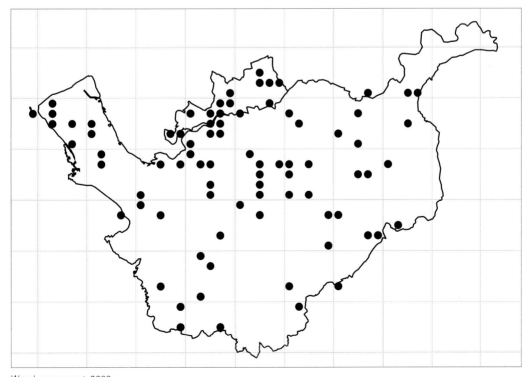

Wood mouse post-2000.

dark brown with white-grey underparts, and can show some yellow to the flanks. Juveniles are greyer and can show a yellow spot on the underside of the throat, so care needs to be taken to avoid confusion with the yellow-necked mouse (*Apodemus flavicollis*) which has yet to be recorded in Cheshire. Older animals can be very pale, often sandy in colour.

Head and body length is 86–103 mm with the length of the tail 75–95 mm. Body weight varies seasonally due to food supplies from 13 to 27 g.

Observation

This animal will readily take artificial food from bird tables and, in cold conditions, will come into houses for warmth and food. The species is more common in the late summer and early winter. In some instances they will readily return to the same mammal trap for food and shelter on regularly trapped sites.

Harvest mouse *Micromys minutus*

Order: *Rodentia*   Family: *Muridae*   Genus: *Micromys*

DAVID QUINN

Status and history

The harvest mouse is a rare but widespread inhabitant of the Cheshire plains, recorded from reed-beds, wet grassland and in the past from arable fields. In 1910, in the absence of firm evidence and lack of records in recent years, Coward tends towards describing the species as very rare. He notes records of harvest mice sightings and nests found in 1893 in Appleton and 1895 near Sandbach.

Some of the records that are held today relate to the introduction scheme co-ordinated by Chester Zoo including records at Chester Zoo, Widnes and Warrington.

In common with many other inhabitants of agricultural land, the intensification of farming methods such as pesticide use and crop rotation is cited as a major cause of the decline in harvest mouse numbers. Linked to this, habitat fragmentation and wetland drainage creates conditions that are difficult for harvest mouse populations undergoing natural 'boom and bust' cycles to overcome. However, the species is often underrecorded and misidentified and little is known of its ecology. Further research to establish the species' true status in the region is required.

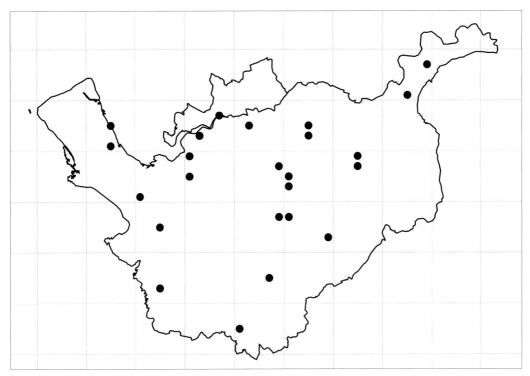

Harvest mouse pre-2000.

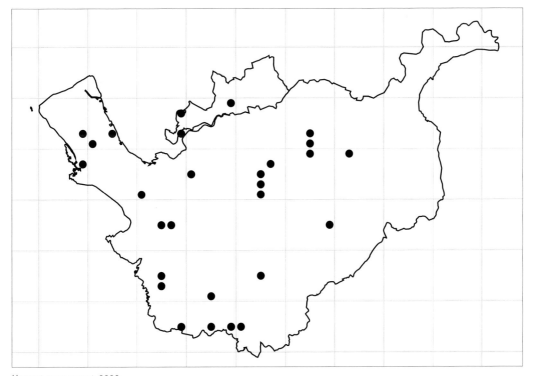

Harvest mouse post-2000.

UK and worldwide distribution

Harvest mice are found throughout central Europe and Asia across to Japan. In the UK the species is found in Yorkshire, Cheshire and all counties southwards, but is absent from most of Wales.

Description

Harvest mice favour areas of tall dense vegetation including long grass, reed-beds, grassy hedgerows, ditches and cereal crops. In Cheshire, more than half of the nests found in 2000 were in habitat described as 'fen' and just under 15 per cent in rank grassland. Harvest mice are extremely active using their prehensile tail as a fifth limb when clambering about in tall vegetation. This is where they build their small nests made of grass 30–60 cm off the ground, split longitudinally to form a ball about 8–10 cm in diameter. Females can breed at 28 days old and in captivity produce up to eight or nine litters of one to seven young; in the wild the figures will undoubtedly be less. Gestation takes 17-19 days and young are born blind and naked. After 11 days the young leave the nest and start to explore their surroundings and are chased away by the female (if she is pregnant) four or five days afterwards.

The maximum life span in the wild is about 18 months, but mortality is high, especially in the first winter. In captivity harvest mice can live for three years.

The natural food of harvest mice is fruit, seeds, berries and some insects. In captivity they have been noted taking aphids from fresh vegetation provided and readily take insectivorous food.

As harvest mice are active 24 hours of the day they have many natural predators including the tawny owl, barn owl, kestrel, pheasant, weasel and stoat.

Key identification features

Harvest mice, along with the pygmy shrew, are the smallest of the British mammals. Head and body length is 50–70 mm with the tail also being 50–70 mm long. Weight is on average 6 g but a pregnant female may weigh up to 15 g. Unlike other mice, the harvest mouse has a blunt nose, small hairy ears and a naked prehensile tail. The general colouration of sexually mature mice is russet orange with clean white underparts. Juveniles are more grey-brown above.

Observation

The simplest way to detect harvest mice is to look for their nests, which is easier in the late autumn/ early winter when the surrounding vegetation has died down.

House mouse *Mus musculus domesticus*

Order: *Rodentia*   Family: *Muridae*   Genus: *Mus*

DAVID QUINN

Status and history

The house mouse is familiar throughout the county although acutely underrecorded. It is considered to be the most widespread terrestrial mammal other than humans. Coward notes that, despite its name, the house mouse is not confined to houses and lives a similar life to the wood mouse.

UK and worldwide distribution

House mice originated in Asia but have since spread throughout the world. The species has been recorded in the UK since at least the Iron Age, and it is widespread throughout the country where there is human habitation, as well as on the inhabited islands.

Description

This species is common in a wide range of urban habitats, including domestic, industrial and commercial properties. Its presence is often given away by a characteristic 'stale' smell and dirty black smears along well-used routes. An excellent sense of balance and ability to run and jump allows house mice to access virtually any area in a property and once established they can reach plague proportions if uncontrolled. Due to their small size they can squeeze through an opening that is small enough to fit a pencil through! If ample food is available they do not move much from the food source and home range can be as small as 5 m². In late autumn and winter, house mice will leave fields and hedgerows and move towards dwellings, a movement that is reversed in the spring.

House mice are rapid breeders and productivity is not affected by daylight length and temperature. In urban situations, a single female can produce up to 10 litters of four to eight young in a year. Gestation takes 19–20 days, and young are weaned at about four weeks old. Young females can reproduce almost immediately after weaning, whilst males are not sexually mature until about 60 days old.

In urban areas house mice will readily eat any food that is available, including cereals, fruit, vegetables and protein products such as cheese. Plaster, soap and glue will also be eaten! In more rural situations their food intake is governed by availability and they will eat almost anything including cereal crops in arable fields, grass and plant seeds in grasslands and even caterpillars, beetle larvae and small insects. They can consume 3–4 g of food each day.

The species is often considered a pest: in agricultural situations it can destroy up to 16 per cent of grain stored. Much of the loss is associated with the necessity to destroy uneaten food that has been contaminated with urine or faeces, or from loss through damaged sacks.

House mice have a number of natural enemies. As with all small rodents, common rats are a frequent predator as is the domestic cat, but birds of prey and predatory mammals will also take house mice.

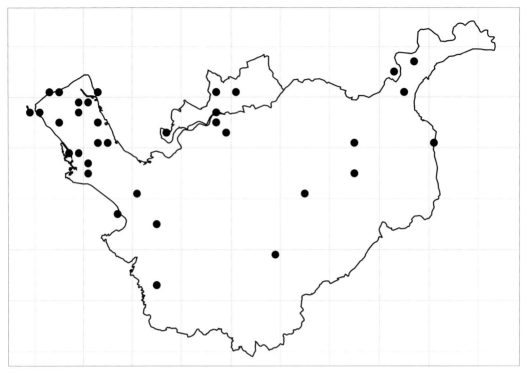

House mouse pre-2000.

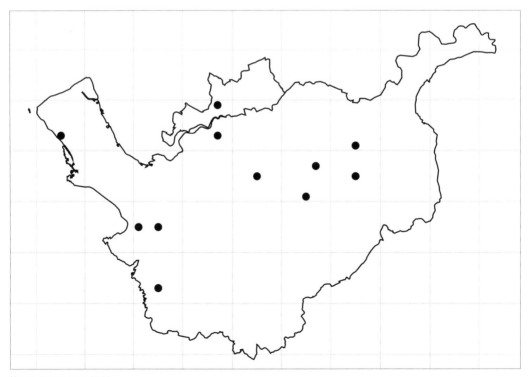

House mouse post-2000.

Key identification features

The head and body length is 76 mm with a long scaly tail measuring 93 mm; body weight varies from 17 to 25 g. They are also distinguished from wood mice (which can also be found in houses) by their greyer appearance (including the underparts, which characterises the subspecies present in the UK), smaller ears and less protruding eyes.

Observation

When inside buildings this animal is easily observed: it is a species that follows regular paths and activities during the course of the day. Out of doors observation becomes more difficult but when spotted the house mouse can usually be approached, as its eyesight is poor. The species is mainly nocturnal, but will venture out in daylight when food is scarce.

Common rat *Rattus norvegicus*

Order: *Rodentia*   Family: *Muridae*   Genus: *Rattus*

DAVID QUINN

Status and history

The reluctance of householders and pest control companies to admit to having met with the common rat is probably the cause for the lack of records of what is perhaps one of the commonest rodents. Coward calls it the most destructive and abundant Cheshire mammal and remarks on how bold they are, feeding on corn thrown down for pheasants, regardless of human presence. This species is widespread and undoubtedly underrecorded in the county.

UK and worldwide distribution

This species is found throughout the British Isles and most of the world with the exception of low altitude inland towns and villages in the tropics and subtropics.

It first appeared in the British Isles in the eighteenth century replacing the native ship rat.

Description

Common rats are adaptable with well-developed senses of hearing and smell. They are typically nocturnal, although they will sometimes forage for food during the day. Normally associated with man, the species occupies a wide variety of habitats, but prefers areas with dense ground cover and a close proximity to water; typically associated with some of the less pleasant locales such as sewers and refuse tips. In optimum conditions where food is plentiful the populations are mostly made up of youngsters, but mortality can be as high as 99 per cent in some circumstances. In the wild, life expectancy is typically 18 months.

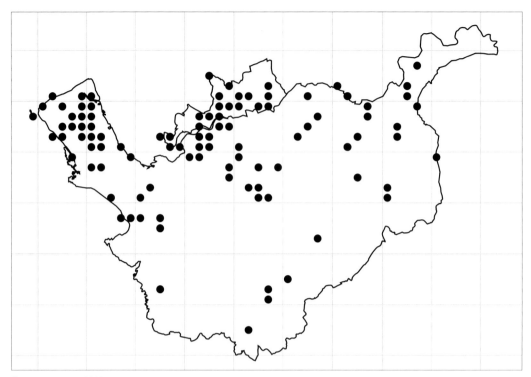

Common rat pre-2000.

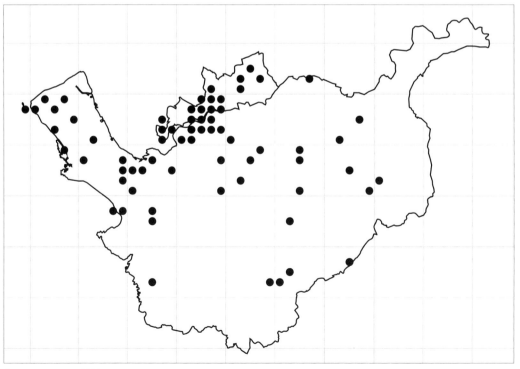

Common rat post-2000.

DAVID QUINN

The common rat (top) and water vole compared.

Rats live in loose colonies, made up of aggregations of clans, usually consisting of a mated pair, or a male and a harem of females. Clans will defend their territory from other rats, and there is a dominance hierarchy whereby the biggest rats are dominant over their smaller counterparts.

Home ranges are generally 50 m in diameter. In areas of high population, fights will often break out involving a vigorous scuffle leading to the loser being chased off and bitten on the rump.

Common rats breed throughout the year. Females can produce up to five litters a year, with larger females capable of producing litters of up to 11 young, although smaller females may only produce five or six. Gestation takes 21–24 days and the young are weaned at about 21 days thereafter. Females can reproduce at about 11 weeks of age.

Although rats prefer starch and protein-rich foods such as cereals, they do have an omnivorous diet that includes meat, fish, vegetables, weeds, earthworms, carrion, crustaceans, nuts and fruit. They sometimes cache food to return to later.

Young rats fall prey to a variety of avian and mammalian predators, but large mature rats deter the smaller predators such as the weasel and little owl. If a population of rats is well established, the domestic cat is unlikely to eliminate the population but can prevent re-establishment after eradication by other methods.

Key identification features

The common rat is unmistakeable for any other frequently encountered rodent with its long scaly tail and pointed muzzle. The fur of the adult is shaggy and grey-brown in colour; young tend to be sleeker and greyer. It can be distinguished from the much rarer ship rat by its larger size, shaggy hair and smaller ears.

Combined head and body length is between 110 mm for a weanling and up to 280 mm for an adult. Tail length is 80–100 per cent of this length at 100–250 mm. Weight varies markedly from 40 g for a weanling to 600 g for an adult male.

Common rats swim well and are sometimes mistaken for water voles.

Observation

The common rat lives where humans reside and can often be seen along waterways or any place where human detritus accumulates. Usually active at dawn and dusk, some immature animals may also be observed during the day.

Ship rat   *Rattus rattus*

Order: *Rodentia*   Family: *Muridae*   Genus: *Rattus*

DAVID QUINN

Status and history

Formerly common around the docks at Birkenhead, species numbers have declined since the arrival of the common rat. The latest record is of one seen in Bromborough in 2007 although the previous record to that sighting was in Northwich in 1996. Coward dramatically stated that the species had, in 1910, vanished from Cheshire and blamed the common rat for murdering them! He did say that a ship rat could be met occasionally at Runcorn and at other ports on the Manchester Ship Canal.

UK and worldwide distribution

Once widespread throughout Britain, the species originates from Asia, and today is widely distributed around the globe. By the 1960s the ship rat was restricted to a few isolated coastal areas in Britain but especially seaports. Since then its range has contracted further, vanishing from Wales, Merseyside and north-east England and it is more or less restricted to dockside warehouses and food stores. Isolated populations still remain on offshore islands such as Lundy. The species is highly dependent upon human habitation, which may make it more vulnerable to pest control measures as it is less adaptable than the common rat.

Description

The ship rat is nocturnal, although it may become more active in the day in undisturbed areas. It lives in groups called 'packs', consisting of several males and two or more dominant females. They are skilled climbers and can also swim well. Nests are constructed from grass and twigs, often in roof spaces.

Less cosmopolitan in its diet than the common rat, the ship rat is an omnivore but tends to prefer plant matter such as fruits and seeds, although it will also feed on insects, carrion, refuse and faeces. On Lundy Island these rats feed on crabs along the shore and on islands they are less of a threat to ground-nesting birds than common rats.

Breeding takes place between March and November. Females can reproduce when they reach a weight of about 90 g (at around three months old) and can produce between three and five litters a year of around seven young. Gestation lasts about 21 days and young are weaned when they are 28 days old. In the wild, life span for a ship rat is up to two years.

In urban situations domestic cats are the main predators of ship rats.

Although associated with the Black Death or bubonic plague, the ship rat is not solely responsible for the spread of the disease worldwide, as the common rat can also act as a carrier.

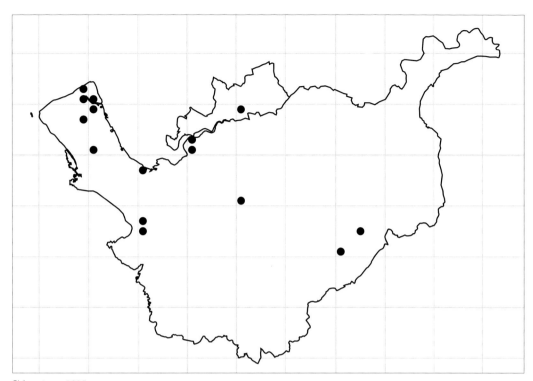

Ship rat pre-2000.

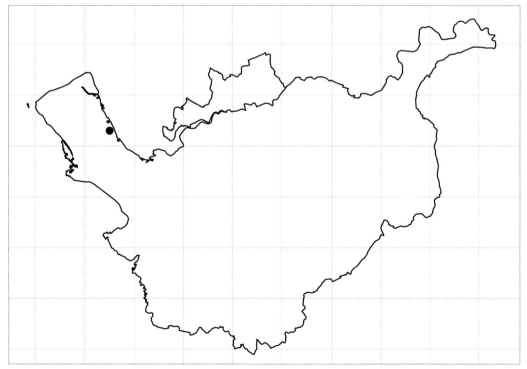

Ship rat post-2000.

Key identification features

Colours range from grey-brown to black above with pale grey or creamy-white colouring underneath. Smaller and tidier in appearance than the common rat, the ship rat has a longer, thinner hairless tail, which is uniformly coloured. Unlike the common rat the ears are virtually hairless. Head and body length is 240 mm; the tail is longer than the body at 280 mm and is used for balance. Body weight can be 150–200 g although males may reach 280 g.

Observation

This species was thought possibly to be extinct in Cheshire, but was recently sighted at Bromborough Docks. The best place to look for a ship rat is probably around the docks at Birkenhead.

5. Rabbits and hares Order Lagomorpha

Rabbits and hares are included in the mammal order *Lagomorpha*, specifically in the family *Leporidae*, and are easily recognisable by their long ears and long hind legs and feet. All lagomorphs are herbivores and have unique double front incisors where the two upper front teeth have a second set behind them. Their nutritional strategy of refection, where soft faeces are produced and eaten, allows maximum nutrients to be gained from their food because it is passed through the gut twice. They also have slit-like noses, which can be open and closed by a fold of skin above.

Three species of lagomorphs inhabit Britain and all can be found in the Cheshire region. Only one of these three, the mountain hare, is truly native to Britain, as the brown hare was introduced by the Romans around 2,000 years ago and the rabbit by the Normans around 800 years ago. In Cheshire the rabbit and brown hare are common but the mountain hare is confined to the Derbyshire borders.

Rabbits and hares in Britain are viewed in a variety of ways: as game, pests and of conservation concern. They have a number of predators including foxes, stoats, polecats, mink, owls and buzzards as well as domestic cats, dogs and man.

Rabbit *Oryctolagus cuniculus*

Order: *Lagomorpha*   Family: *Leporidae*   Genus: *Oryctolagus*

DAVID QUINN

Status and history

First introduced to Britain by the Normans as a food source, rabbits are now common and widespread although they have suffered huge reductions in the past from the myxomatosis virus. Coward marks the species as 'everywhere abundant' except in the immediate neighbourhood of towns and over altitudes of 1,700 ft (500 m). In Cheshire today they are common and considered pests in some areas because of the vegetation they eat and the damage caused by their burrows.

The smallest of the lagomorphs in Cheshire, rabbits are usually greyish brown and have smaller ears than the hare. Black varieties sporadically occurred in the region in the past especially in Lyme Park. Small colonies live at various sites throughout the county including at Pickerings Pasture Local Nature Reserve in Widnes, Griffiths Road Lagoons in Northwich and at Moore Nature Reserve in Warrington. The name 'clargyman' was given to these black rabbits in Cheshire because the clergy used to wear collars made of black rabbit fur. Black rabbits tend to be smaller

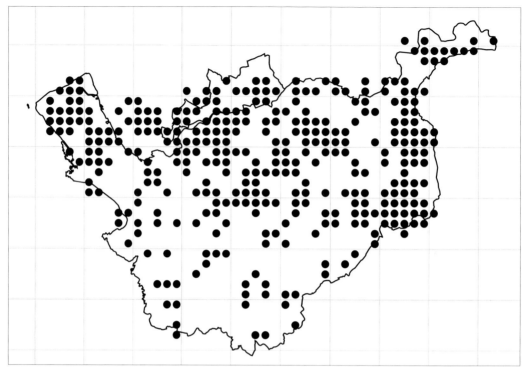

Rabbit pre-2000.

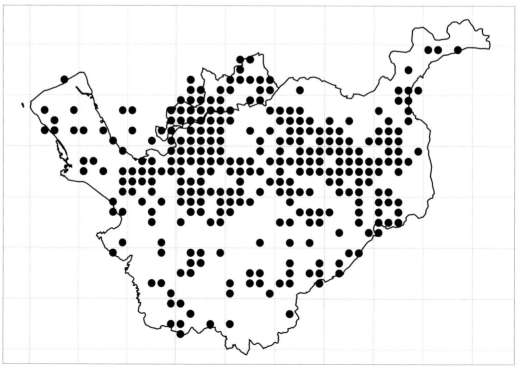

Rabbit post-2000.

than normal specimens and rarely live to become adults.

UK and worldwide distribution

Rabbits are widespread in western Europe, including the Balearic Islands, Corsica, Sardinia, Sicily and the British Isles. They are also found in North Africa and have been introduced to Australia, New Zealand and North and South America.

Description

Rabbits are found in a variety of habitats including grassland, heathland, moorland, scrub, open meadow, dry sandy soil, woodland and agricultural crops. They avoid coniferous forests. Life span is up to nine years.

The species is famed for being a prolific breeder and this has certainly helped with its colonisation of Britain. The breeding season starts in late January and continues until August; rabbits have between three to seven litters per year and an average of five young in each litter. Mothers adopt an 'absentee' strategy to rearing the litters, visiting to suckle the young only once a day. It is thought this is the best strategy for the mother to take in order to avoid predators detecting the litter.

Rabbits eat the leaves of a wide range of vegetation including agricultural crops, cereals, young trees and cabbages. In winter, they eat grasses, bulbs and bark. They reingest their faeces for nutritional benefit.

The burrows excavated by rabbits are known as warrens. Tunnels can be 1-2 m long, with the nest at the end lined with grass, moss and belly fur. Rabbits use regular trails, which they scent mark with faecal pellets. They live in social colonies with a dominance hierarchy; several dozen rabbits may inhabit a large warren. Rabbits spend the daytime underground in their burrows and are mainly active outside at night.

Key identification features

Rabbits are smaller and less gangly than hares, and have shorter ears. The body fur is brown-grey, the tips of the ears are brown, and the upper surface of the tail is dark brown. The characteristic white flash on the underside of the tail can be seen when the animal is fleeing. Head and body length is 30–40 cm and weight is 1.2–2 kg.

Observation

Their burrow entrances and droppings are a good indication of their presence in an area. You need not look hard to see rabbits in Cheshire!

Brown hare   *Lepus europaeus*

Order: *Lagomorpha*   Family: *Leporidae*   Genus: *Lepus*

DAVID QUINN

Status and history

Brown hares were introduced to Britain by the Romans probably from Asia around 2,000 years ago and became widespread throughout lowland England and Scotland. As brown hares are hunted, it is possible to gain a great deal of information about their distribution and abundance through game bag records which showed a fall in hare numbers of around 80 per cent in the twentieth century. Coward noted that the species had 'been reduced in numbers in many districts since the passing of the Ground Game Act in 1880'.

Brown hares suffered noticeable population declines in the 1960s and 1970s due to a combination of factors. These include the widespread intensification of agricultural practices, such as the conversion of grassland to arable crops, and changes in cropping regimes, which may remove important food sources at vital times of the year. Shooting, poaching and coursing are likely to have contributed to the decline, as has the increase in the numbers of the hare's major predator, the fox. Hares are now more common in arable than pastural areas.

In Cheshire, brown hares suffer from habitat fragmentation. A questionnaire undertaken in 2000 indicated that the hare population in Cheshire was around 6,133 individuals.

UK and worldwide distribution

Brown hares are widespread in central and western Europe including England and Wales, but they are absent from north-west Scotland, Sardinia, the Balearic Islands and most of Spain and Portugal. They were introduced to Ireland for sport in the nineteenth century where their spread has been checked by competition from the Irish hare, a subspecies of mountain hare.

Description

Brown hares prefer temperate open habitats. They are found in most flat country among open grassland and arable farms and use woodland and hedgerows as resting areas in the day. Their diet consists mainly of herbs in the summer, and predominantly grasses in the winter. They also feed on cereal and root crops and, in bad seasons, it has been suggested that they may eat animal corpses.

Unlike the rabbit, the brown hare does not excavate burrows but instead rests in a shallow depression known as a form, where only its back and head are visible. The form may be against a hedge, in short grass, scrub or a ploughed furrow and, when lying in its form with ears laid flat, a hare is well camouflaged. They are mainly nocturnal and solitary with an adult occupying a range of 300 hectares, which it may share with other hares as they are not territorially aggressive.

The breeding season lasts from mid-February until mid-September. Courtship involves boxing, the traditional 'mad March hare' behaviour, which is actually an unreceptive female fending off a male rather than fighting between males. Females have one to four litters per season with one to four young in each litter. Because they do not have the safety of a burrow as rabbits do, baby hares, known as leverets, are born with fur, their eyes open and active.

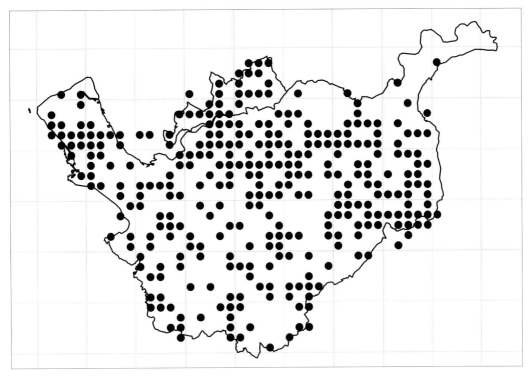

Brown hare pre-2000.

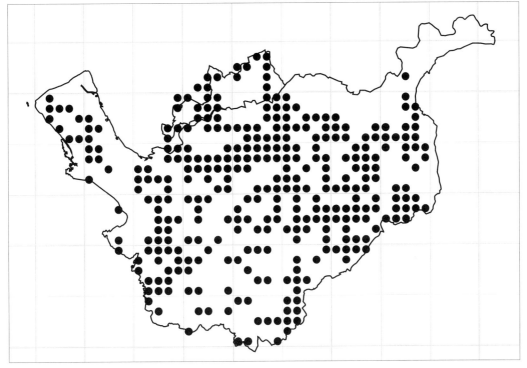

Brown hare post-2000.

Brown hares are the fastest land animals in the UK and, with their powerful hind legs, they escape predators by out-running them at speeds of up to 70 kph. Life span is up to 13 years. Foxes, stoats, buzzards and owls may all eat leverets.

autumn, the summer coat being a little lighter than the reddish winter coat. The tail is black on the upper surface and white underneath. In contrast to rabbits, which have a brown iris, the brown hare has a golden iris and a black pupil.

Key identification features

The general form and structure of the brown hare resembles that of the rabbit, but obvious differences include the hare's longer, larger body, much longer hind legs, and longer ears with black tips that are equal in length to the head. The hare has a loping gait and the tail is held down when running showing its black top. Body and head length is 52–60 cm with weight 3–4 kg.

Generally, brown hares are a brown-russet colour, with a white underside. The fur moults in spring and

Observation

The best time to see a hare is in early morning or at dusk, when it is feeding. In between nibbling plants, it frequently sits up to have a good look around, its keen sense of smell and hearing helping it to detect predators. When moving around a field, the hare stays close to the ground with its ears flat along its back, travelling slowly and carefully to remain inconspicuous. When threatened, however, look out for the black-topped tail as the hare zigzags across the field at speed.

Mountain hare *Lepus timidus*

Order: *Lagomorpha*   Family: *Leporidae*   Genus: *Lepus*

DAVID QUINN

Status and history

Mountain hares have limited distribution in the United Kingdom. They are found in the Highlands, Borders and south-west of Scotland, the Isle of Man, the Peak District and Ireland. Although the only truly native lagomorph to Britain, they would not exist in this region without the help of man. Introduced

to Yorkshire from Perthshire for hunting in about 1880, mountain hares expanded into Derbyshire and, by 1908 if not earlier, had crossed the border into Cheshire.

Coward referred to the mountain hare as the 'varying hare' with its local name of 'white hare'. He stated that the species had firmly established itself on the Cheshire uplands in the north-east of the county

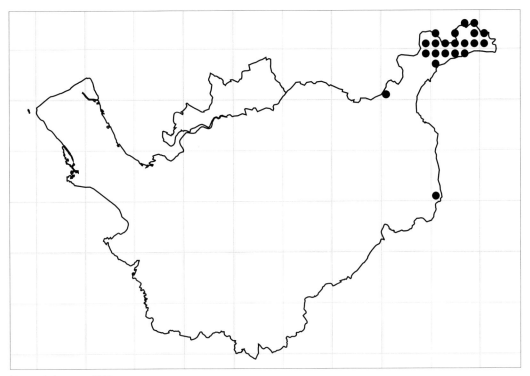

Mountain hare pre-2000.

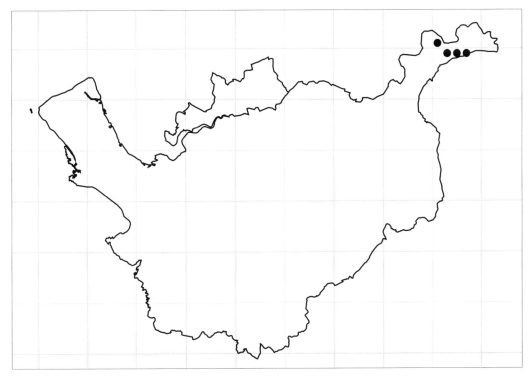

Mountain hare post-2000.

and in 1893 Colonel Lees' gamekeeper counted no fewer than 50 hares with his field glasses.

Today in Cheshire they are still restricted to the hills of the north-east of the county.

UK and worldwide distribution

Mountain hares are found across northern Europe, including northern England, Ireland (where it is known as the Irish hare), Scotland and also the Alps, where they live at altitudes of up to 1,300 m. They are also found across northern Asia, Alaska and Canada.

Description

Mountain hares, as the name suggests, are mainly found at high altitudes although in places where brown hares are absent they are also found in lower grassland areas. They live on heather moorland, dry rocky hilltops and occasionally woodland up to the snowline where their diet consists of young heather, grasses, herbs, sedges, rushes and bilberry. Mountain hares have a range of 80–100 hectares.

Mountain hares are mainly solitary, but in severe weather, or at good food sites, may congregate in large groups of up to 70 animals. Similar to the brown hare, they live in a form or under the cover of heather or rock outcrops although on occasion they may burrow for shelter, particularly in deep snow. Mountain hares can reach speeds of over 60 kph when threatened.

They are mainly nocturnal but may be more active during the day in the mating season, which lasts from February to August. Females have one to three litters per year with between one and five leverets in each litter. Gestation is 50 days; the young are born with fur and their eyes open, and are weaned at three weeks. Their greatest danger is from predation by red foxes, but stoats also prey on large numbers of leverets.

The mountain hare is also known as the blue hare due to one of its summer colour forms.

Key identification features

Mountain hares moult three times a year and in the summer they are dusky brown on top and grey-blue underneath. In the winter they have striking white coats, the extent of the white colouring being related to temperature. They are smaller than the brown hare but larger than the rabbit, and have a more rounded body shape and no black upper surface on the tail although they do have black ear tips. Mountain hares have shorter ears and legs than brown hares.

Life span is up to 10 years. Head and body length is 50–60 cm with a weight of 2.5–4 kg.

Observation

Seeing a mountain hare is not easy as they are largely nocturnal and their late spring moult makes them look similar to brown hares. One of the best times of year to see them is in the early spring when there has been a snow melt and their white winter coats give them away on the open hillsides. However, if there are a few patches of snow left they will rest on these areas which easily avoids them being seen. During periods of snow cover they gather on leeward hill slopes in groups to shelter or feed where shallow snow permits scraping to reveal underlying heather.

Raised peat bog, Risley Moss.

ROB SMITH

Stockton Dingle, Malpas.

SUE TATMAN

Conifer woodland, Daresbury. ROB SMITH

Moorland, east Cheshire. BEN HALL

Cheshire farmland, Peckforton. SARAH BIRD

Mossland, Holcroft Moss.   KAT WALSH

Sankey Brook, Warrington.   ROB SMITH

Wildlife Garden, Chester Zoo.   SARAH BIRD

Common reed-bed, Astmoor, Runcorn.   PAUL OLDFIELD

Cheshire's industrial landscape, Helsby.

Grey squirrel.
ROB SMITH

Red squirrel.
ROB SMITH

Hazel dormice.

Bank vole.

Northern water vole.

Wood mouse.

House mouse.

Brown hare.

Mountain hare in winter pelage.

Hedgehog.   ROB SMITH

Brown long-eared bat.
ANDY HARMER

Fox.

Badgers.

Polecat.

Otter.

Grey seal.

Common dolphin.

Red deer
stags and hind.
ROB SMITH

Roe deer stag.

MIKE ROBERTS

Fallow deer buck shedding velvet.

Fallow deer hind and calf.

6. Insectivores
Orders Erinaceomorpha and Soricomorpha

These orders are among the oldest and largest surviving mammalian lineages. The insectivores are considered to be the most primitive of living placental mammals and still have much in common with their earliest ancestors the *Zalambdolestes*, from the Cretaceous period. Primitive characteristics include walking on the soles of their feet (plantigrade) and possessing a cloaca: a combined intestinal, urinary and genital tract. Their primitive dentition does not lend itself easily to dental formulae and it is hard to distinguish canines from incisors and molars.

All insectivores are small; none is larger than the rabbit. All have poor sight and rely heavily on their sense of smell, which is very sensitive owing to the relatively large olfactory centre in the brain.

A few insectivores, however, possess highly developed characteristics such as anti-predator spines or poisonous salivary glands.

Out of the six families that exist, three can be found in Europe. These are geographically the most widespread and include *Erinaceidae* (hedgehogs), *Soricidae* (shrews) and *Talpidae* (moles).

Five species are found in Cheshire; although all are considered common, their largely nocturnal lifestyle and shy behaviour mean that they are difficult to study. Molehills are the most obvious evidence of mole presence whereas shrews are regularly found during small mammal trapping sessions and hedgehogs are regular visitors to gardens.

Hedgehog *Erinaceus europaeus*

Order: *Erinaceomorpha*   Family: *Erinaceidae*   Genus: *Erinaceus*

DAVID QUINN

Status and history

The hedgehog is one of our most easily recognised mammals. Hedgehogs have always been common and widespread in Cheshire, although not always as popular as today. In past centuries they were persecuted for taking the eggs of ground-nesting game birds, sometimes with a bounty on their heads. Coward

notes that in 1656 the churchwardens of Bunbury paid bounties on 253 animals, and at Rostherne in 1673 and 1674 rewards were paid for 316 hedgehogs. In 1896 in Cranage, 60 hedgehogs were recorded on a gamekeeper's gibbet. These figures indicate that, despite persecution, hedgehogs remained numerous. Hedgehogs also suffered a form of literary persecution from another Cheshire resident, Lewis Carroll:

in *Alice's Adventures in Wonderland* the unfortunate animals were used as balls in a game of croquet.

Nationally, it is difficult to estimate hedgehog population size, but they are generally believed to be in decline. Sadly the best way to estimate numbers is to count road casualties. Although hedgehogs are very vulnerable to collision with motor vehicles, it is not clear if death on roads is a significant factor limiting their numbers. What is apparent is that Cheshire has a much lower population density than many other parts of the country. This is probably due to our high badger population: badgers are significant predators of hedgehogs as they are able to open a rolled-up hedgehog with their powerful claws. A large proportion of Cheshire hedgehog records are from domestic gardens, often in relatively built-up areas. As such places are unlikely to be frequented by badgers, it appears that they provide a very important refuge for Cheshire's hedgehogs.

UK and worldwide distribution

Hedgehogs occur in all parts of the UK and Ireland. They have been introduced to several offshore islands, including the Outer Hebrides, where they spread rapidly and have done great damage to breeding wader populations.

The hedgehog is found across western Europe, through France and Germany, south into Spain and Italy, and to the north-east into southern Scandinavia and the western Russian states. Further east, the eastern hedgehog (*Erinaceus concolor*) replaces it. Hedgehogs have been introduced to New Zealand, where they are now common in lowland areas.

Description

Hedgehogs are solitary animals, mainly active at night. Each animal has a home range, which may overlap those of several other hedgehogs, of both sexes. Although several hedgehogs may use an area, they generally avoid each other.

The natural habitat of the hedgehog is woodland and woodland edge, with plenty of dense undergrowth. Hedges, especially with untidy vegetation at the bottom, are also good habitat. They have adapted well to domestic gardens and the abundance of hedges, shrubberies and overgrown corners suits them well.

Hedgehogs will spend the day in a nest, which is made from a tightly compacted mass of dead leaves, with a central chamber just large enough for the animal. Nests are occasionally built underground, in a crevice or burrow, but more usually at ground level. They are often partly sheltered by undergrowth,

against a fallen tree, or even under a woodpile. Nests have even been found in the gaps under garden sheds. Each hedgehog has several nests, scattered around its home range, and will change its resting place every few nights. Breeding nests are similar, but larger to accommodate a growing family.

Hedgehogs hibernate in similar nests and will rouse several times during the winter and often move between nests, even building a new one before resuming hibernation.

Although beloved by gardeners for eating slugs, hedgehogs will take a wide variety of invertebrate prey including beetles, caterpillars, earthworms, small snails, spiders, millipedes and woodlice. If available they will take birds' eggs and carrion and can travel up to 4 km while foraging each night, but a few hundred metres is more usual. Although hedgehogs usually appear to hug the ground, their legs are surprisingly long and when they want to they can move rapidly. They have poor eyesight but an excellent sense of smell, and very good hearing.

Many theories have been put forward for the hedgehog's bizarre behaviour of self-anointing, but the explanation is still unknown. The behaviour is usually associated with some strong-smelling substance: a huge range of triggers has been reported. The hedgehog sniffs, licks and chews the substance, and produces mouthfuls of frothy saliva. It then spits and licks the saliva over its shoulders and flanks. This behaviour can occur for a few minutes or several hours, it may be done by either sex and at any time of year.

The breeding season starts in May, shortly after the females emerge from hibernation, and lasts through summer. Generally a female only rears one litter per year, although if she loses an early litter she may mate again. Courtship is a noisy affair, and the female is initially aggressive towards the male, snorting and butting him with her head. The male tries to get behind the female to mount her, and may end up chasing her in circles, sometimes for several hours. Mating is presumably uncomfortable for the male, as the female does not flatten her spines for him, as is sometimes reported. There is no truth in Aristotle's theory that hedgehogs mate belly to belly, with the female lying on her back! The male takes no part in raising the young.

Pregnancy is 31-39 days; in Britain average litter size is four to five. The young are born blind, pink, hairless and with no spines. About a hundred spines are present in the skin of the back and within 24 hours of birth these emerge and harden. The young grow rapidly so that at two weeks old the eyes and ears open, and they can roll up defensively. By four weeks they are becoming more active and start to

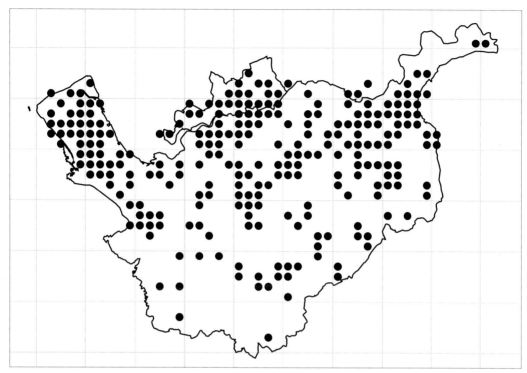

Hedgehog pre-2000.

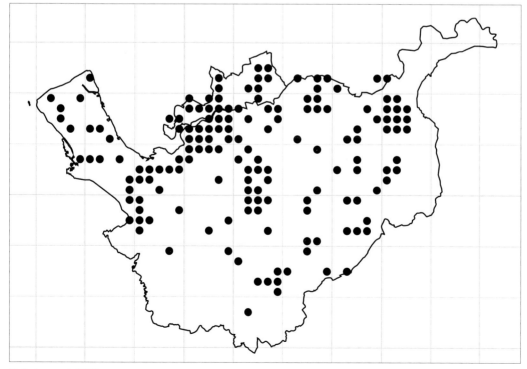

Hedgehog post-2000.

accompany their mother on foraging trips. They are weaned at five to six weeks and become independent soon afterwards. They have the rest of the summer to disperse and grow large enough to survive their first winter, for which they need to weigh at least 450 g to survive hibernation. Although hedgehogs can live up to eight years in the wild, life expectancy at weaning is about two years.

The hedgehog's main food source of invertebrates is largely unavailable in winter and hibernation is their strategy to survive the period when food is scarce. They enter hibernation in October or November and emerge in March or April, the exact time depending on climate, individual body condition and sex. Males tend to start and end hibernation earlier than females because they are free to accumulate fat throughout the summer (when females are suckling young). During hibernation the metabolism slows right down and the body temperature drops. Heartbeat and breathing also slow. Energy requirements are very low in this state, and the hedgehog can survive on stores of body fat accumulated in the autumn. Even so, their body weight can drop by a third before spring. Hedgehogs will wake up many times during hibernation, on average rousing every seven to 11 days, although they do not always leave the nest. Hedgehogs found active in winter should not be considered in need of assistance, unless they are clearly underweight.

Hedgehogs face many hazards, some found in gardens, others in the wider countryside. When trying to drink they can fall into ponds and swimming pools and drown; gently sloping sides around part of the pond perimeter allow them to climb out easily. Although hedgehogs are good climbers capable of scaling rough garden walls, they cannot manage the smooth vertical sides of the pit under a cattle grid; a ramp of wood or rock in one corner will enable them to escape. Hedgehogs have also been found caught up in tin cans and the plastic mesh used to hold beer cans together. As well as removing a food source, slug pellets may be toxic to the hedgehog. Strimmers can cause horrific injuries and many hedgehogs can get tangled in garden netting, and can starve if not found

and released. Every year many hedgehogs get roasted when they decide a bonfire heap is an ideal nesting site, so the base of a fire should be checked before lighting, or the fire built on a clear base. Rolling into a spiky ball is no defence against car tyres and motorists should keep their eyes open when driving at night.

Key identification features

Hedgehogs are our largest British insectivore, with a head and body length of 200–300 mm. Adults weigh around 600–700 g in summer; males tend to be slightly larger than females. Before winter hibernation the accumulation of body fat means that they can weigh 900–1,200 g.

The hedgehog's most obvious feature is the spines covering all of the upper body except for the face, the rest of the body being covered in sparse coarse hairs. The spines are specially adapted hairs, and an adult hedgehog has about 5,000 of them. When threatened, the hedgehog rolls up and uses strong muscles beneath the skin to pull the spine-bearing skin on the back around so the face and soft underside are completely covered, forming a tight prickly ball. It will stay tightly rolled until it feels safe enough to uncurl. Folklore accounts of foxes forcing hedgehogs to uncurl by rolling them into water or even urinating on them are unlikely to be true. Only badgers can easily unroll a curled hedgehog and, when they do so, will consume the whole animal leaving only the spine-covered skin.

Observation

Gardens can be ideal habitat for hedgehogs, and are often the best place to observe them. Garden hedgehogs can be attracted into view by a regular supply of suitable food such as canned dog or cat food, minced meat or scrambled egg. Special hedgehog food is also now available. Hedgehogs should not be offered bread and milk as they cannot digest cow's milk and it will give them a nasty stomach upset.

European mole *Talpa europaea*

Order: *Erinaceomorpha*   Family: *Talpidae*   Genus: *Talpa*

DAVID QUINN

Status and history

The abundance of the mole has been well documented in Cheshire since the early 1700s. Historically they were considered to be pests and trapped on a huge scale with their pelts used for clothing: in 1732 the churchwardens of Prestbury paid for 5,480 moles to be killed. Coward notes that they are abundant throughout the county. Lack of recording pre-2000, noticeable with many of our more common species, gives an unknown distribution and population, but post-2000 there has been a revival in mole recording. Although there are fewer records from the south-west of the county and from the Wirral, existing records generally show a widespread distribution of the species across the county.

UK and worldwide distribution

Distributed widely throughout mainland Britain, moles are missing from Ireland and most islands, with the exception of Jersey, the Isle of Wight, Anglesey, Mull, Alderney and Skye. They are widespread in Europe except for the more southern countries and present throughout Scandinavia and western Russia.

Description

Moles need to be able to construct an underground network of tunnels and semi-permanent burrows to exist. Thus they avoid stony, waterlogged and sandy soil where they cannot construct a proper burrow system. Favoured habitats with soil deep enough to allow tunnelling include arable fields, deciduous woodland, permanent pasture and lawned parks and gardens. They do not commonly occur in coniferous forests, sand dunes or moorland, where invertebrate prey, especially earthworms, is scarce.

Moles move about in a network of 'galleries' about 15 cm below the surface but which extend down to 50 cm. Part of this network is semi-permanent, used by successive generations of moles and tending not to have molehills as the layout is old and there is no more earth to excavate. Temporary 'hunting galleries' are dug out, exploited only by a single individual and not reused. When such hunting galleries are excavated the mole expels the debris in the form of molehills. The whole of the network has a length of 100–200 m; it also comprises a resting place including a nest of dried leaves situated at the junction of several galleries, as well as secondary chambers, at times stocked with food reserves. The speed of movement of a mole in a gallery is quick at about 1 m per second. The burrowing capacity is about 20 m per day.

The European mole is entirely carnivorous: it consumes about its own body weight of food each

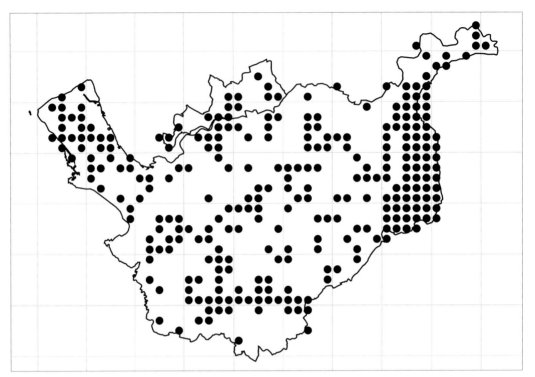

Mole pre-2000.

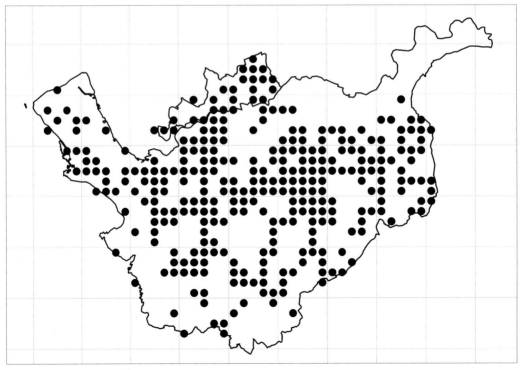

Mole post-2000.

day. Moles are active day and night feeding mainly on earthworms, though they will take beetles, fly larvae, myriapods, molluscs and occasionally amphibia. They have tiny eyes and ears as the majority of their time is spent underground in the dark. Their tactile senses are extremely well developed; the sensitive hairs on the snout, feet and tail help moles to seek out their prey, as do their large spade-like feet, which also prove invaluable for digging.

Moles are typically solitary, and both sexes defend their territories vigorously. Males extend their tunnel systems during the short breeding season as they search for females. They breed from late February to June. Usually a single litter per year is born of between two and seven naked, blind young. The young are suckled for about a month and leave the nest at around 33 days old to disperse above ground. This is a dangerous period as they are extremely vulnerable to predators including owls, buzzards, stoats, dogs and cats. They become independent at eight months and are able to breed within 11 months.

Female moles are the only mammals known to posses reproductive organs called 'ovotestes', which contain a normal functioning ovary as well as a testicular area that produces a large amount of testosterone. This intriguing feature may explain why female moles are as aggressive as males when defending their territories; it may also account for the external similarities between males and females.

Main predators of moles are owls, buzzards, herons, foxes and stoats. Moles are often killed by humans by trapping or poisoning which can result in a slow and painful death. They have no protection in law and are considered to be pests by many, although there are doubts as to whether this perception is correct.

Key identification features

Moles have an elongated, cylindrical body covered in short black, velvety fur which can lay in any direction, thus helping with movement through the tunnels. Their eyes are tiny and often hidden by fur and they have no external ears. The large forepaws have five robust claws. Moles weigh 65–130 g and measure up to 16 cm in length with up to a further 4 cm for the tail. Although they can live for up to seven years, the average life span is three years.

Observation

Moles are rarely seen above ground and evidence of their presence is found in the form of fresh molehills.

Common shrew *Sorex araneus*

Order: *Soricomorpha*   Family: *Soricidae*   Genus: *Sorex*

DAVID QUINN

Status and history

The common shrew has been recorded throughout the majority of Cheshire since the late 1800s. Coward describes the species as abundant and also comments on the considerable altitude at which the shrew can be found and the fact of seeing the species' footprints in the snow. Records between 1980 and 2000 show a wide coverage of common shrews across the county although after 2000 records are mainly confined to the north of Cheshire, which may indicate a fall in numbers but could equally be related to a simple lack of recording.

UK and worldwide distribution

The common shrew is distributed widely throughout central and northern Europe. With the exception of Britain it is largely absent from western Europe as well as from Ireland, the Isle of Man, Northern Isles and the Outer Hebrides. In Jersey, it is replaced by the French shrew (*Sorex coronatus*).

Description

The common shrew can be found throughout Cheshire in a variety of habitats. They prefer dense grassy 'edge' habitats such as field margins and road verges, though they can be found in hedgerows, deciduous woodland and scrub. They build nests below the ground or under dense vegetation. Common shrews are active both day and night (but especially at night), and rest for only a few minutes between bouts of activity. They spend much of their time under leaf litter, or similar,

searching for food, or using old mouse runs to get around.

Except for when rearing young, shrews are solitary and are extremely aggressive towards each other. They defend territories, which vary in size from 370 to 630 m² and usually last the shrew's lifetime. Males extend their ranges during the breeding season in their search for females.

Shrews have the ability to echolocate and have a good sense of smell and hearing, but their eyesight is poor; they locate prey hidden up to 12 cm deep in soil by probing and sniffing with their snout. They feed mainly on invertebrates, including earthworms, slugs, spiders, beetles, snails and woodlice. They need

DAVID QUINN

The pygmy (top), common (centre) and water shrew compared.

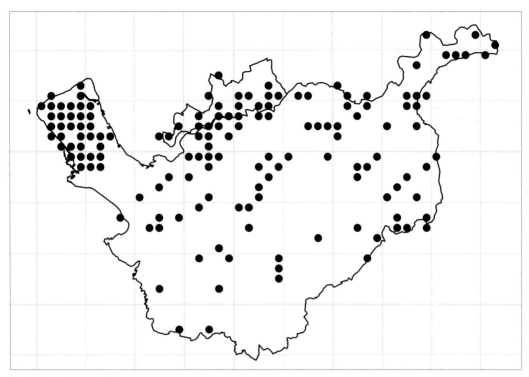

Common shrew pre-2000.

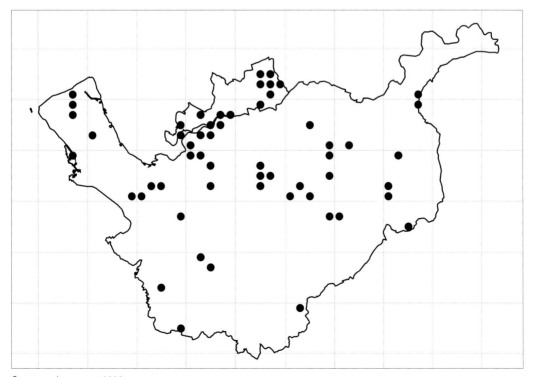

Common shrew post-2000.

to consume 80–90 per cent of their body weight in food each day and, as such, the young suffer high mortality rates with 50 per cent dying within the first two months of life. The life span of a common shrew is a maximum of 23 months. Mating takes place between April and August and animals are sexually mature in the spring following their birth. Females raise two litters per year with an average litter size of six to seven young. When disturbed from the nest, young common shrews will sometimes follow their mother in a caravan fashion, using their mouths to hold on to the tail of the sibling in front.

Like hedgehogs, common shrews self-anoint: the animal sniffs, licks and chews a usually strong-smelling substance to produce mouthfuls of frothy saliva which it then spits and licks over itself.

The main predators are owls, but weasels, stoats and foxes also hunt shrews. Domestic cats often kill them but they are apparently distasteful and are rarely eaten. All shrew species are protected in the UK by the Wildlife and Countryside Act 1981.

Key identification features

Common shrews weigh 5–15 g and are up to 5.5 cm in length. They can be distinguished from other shrews by their tri-tonal colouration: dark brown back, lighter brown flanks and grey-white belly. They have a long snout and whiskers, which constantly probe for food. Occasionally, individuals can be found with white spotting on the ears and a white tail tip. White spotting throughout the fur is also fairly common.

Observation

Shrews are not easily seen because they move very quickly, but they can often be heard, particularly in March and April when they meet to mate. They communicate by a series of high-pitched shrieks and chatters, which are particularly loud if the shrew is alarmed or angry. As one of their main predators is the domestic cat, shrews are unfortunately often observed being brought into houses, dead or alive.

Pygmy shrew *Sorex minutus*

Order: *Soricomorpha*   Family: *Soricidae*   Genus: *Sorex*

DAVID QUINN

Status and history

The pygmy shrew was first recorded in Cheshire in 1894 when one was caught by a cat in Rainow. Coward refers to the animal as the 'Lesser Shrew' and surmises the species 'is not so uncommon in Cheshire as is generally supposed, but is certainly not so abundant'.

According to records, pygmy shrews seem to have occurred more frequently in the north of the county

before 1980. After 1980 there are fewer records, although those records seem to be patchily distributed countywide.

UK and worldwide distribution

The pygmy shrew is distributed widely throughout Britain and Europe, though they are absent from the

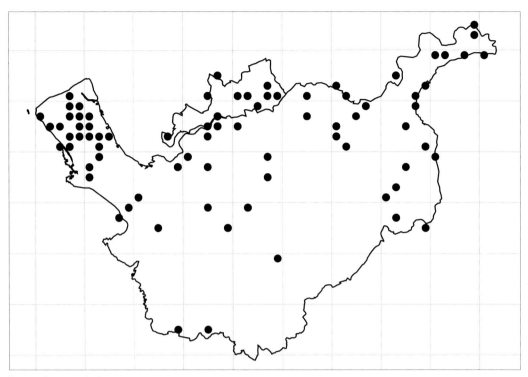

Pygmy shrew pre-2000.

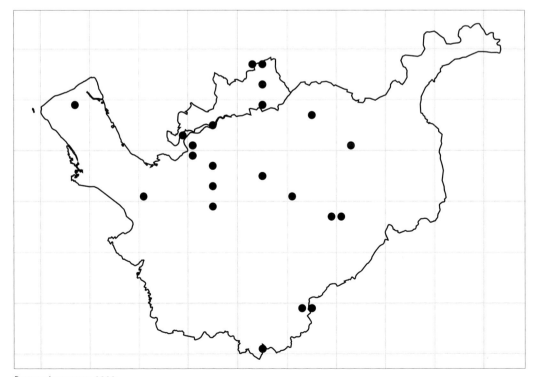

Pygmy shrew post-2000.

Shetland, Scilly and Channel Isles. They are the only species of shrew to be found in Ireland. On continental Europe they are also widespread, although they are absent from some southern areas.

Description

The pygmy shrew occurs in a very broad range of terrestrial habitats wherever there is adequate ground cover, such as grassland, heathland and sand dunes, although they tend to be absent from woodland. They rarely live in underground burrows, preferring longer vegetation for cover and hunting.

Pygmy shrews are active by day and night, interspersing bouts of activity with rest periods. They are typically solitary, and will defend their range against other pygmy shrews. Unlike common shrews, pygmy shrews do not dig for food, instead making surface tunnels through vegetation and feeding on invertebrates such as beetles, spiders and woodlice that can be found in the leaf-litter. Nor do they compete with common shrews for earthworms, possibly because this prey is too large for them to handle. Pygmy shrews are well known for their voracious appetites: due to their small size and high metabolic rate, they have to eat regularly, and need to consume about one-and-a-quarter times their body weight each day to survive. They do not hibernate, as they are too small to store the necessary fat reserves, so instead they have to remain active during winter.

Births occur between April and August, peaking in June. Two litters are usually produced each year, each consisting of four to seven young. The young overwinter as immatures, reaching sexual maturity the following year, although some females born early in the year may actually breed in the year of birth. Main predators of pygmy shrews are owls, raptors, mustelids, foxes and cats, though pygmy shrews try to intimidate their predators by secreting foul-smelling oils, which taste unpleasant. The maximum life span is 13 months.

Key identification features

As both the common and scientific names suggest (*minutus* means small), the pygmy shrew is tiny and is in fact the smallest native British shrew. They weigh 2.5–5 g and are up to 65 mm in length. The coat is bi-coloured with a grey-brown back and a pale belly. Compared to other species of British shrews, the pygmy shrew has a relatively longer, hairier tail. They have a long pointed snout and whiskers, which constantly probe for food. Their teeth are red tipped, formed by the deposition of iron, which toughens them against wear and tear.

Observation

Pygmy shrews are the most common mammals to be found caught in discarded bottles and drink cans. Tipping out the contents of these containers may reveal parts of the shrew skeleton. Like the common shrew, the pygmy is often a victim of domestic cats.

Water shrew *Neomys fodiens*

Order: *Soricomorpha* Family: *Soricidae* Genus: *Neomys*

DAVID QUINN

Status and history

The water shrew, which is locally named the otter shrew, was widespread throughout Cheshire in the late 1800s to the early 1900s. Coward notes the species is common in brooks and ditches in all parts of the county and he comments on its seeming indifference to the speed of water in which it lives.

Before 1980, species records were confined to the Wirral and south and east Cheshire. Post-1980 records for the water shrew are largely spread across the middle of the county, with a slight northerly bias. The species is uncommon in the county today but although it is the least abundant and least widespread of the British shrews, it is not considered rare. In some areas it seems to have declined drastically in number; this is probably due to the destruction of its habitat by the draining of waterways and wetlands and also pollution. Sometimes the water shrew is regarded as a pest because it eats the spawn of valuable fish stocks.

UK and worldwide distribution

Water shrews are widely distributed throughout mainland Britain, though they are generally uncommon. They are scarce in the Scottish Highlands, are found only on the larger Scottish islands and are absent from Ireland. They are also to be found in most European countries and across central Asia.

Description

Water shrews have a semi-aquatic lifestyle and are generally found in habitats close to water, such as wet grassland, fens, reed-beds, streams, pond margins, on riverbanks and in ditch systems. They can even be seen by the seashore. Occasionally water shrews can be found in rough grassland, woodland and scrub when they are dispersing. They have been found up to 3 km from a water source.

Mostly active at night, during the day water shrews remain in a ball of vegetation in the extensive underground burrow systems that they have excavated. In the water, a water shrew has to paddle fast to stop itself bobbing up to the surface as air trapped under its fur makes it buoyant. A fringe of bristly hairs along the underside of the tail acts as a rudder; the tail itself can be used to grip twigs and branches. It hunts along the bottom of a watercourse, turning over stones with its grasping feet, and can stay underwater for 20 seconds at a time. The water shrew does not like being in the water for long periods and will dry its fur by squeezing along the narrow passageways of its tunnel and then grooming itself thoroughly.

The shrew's mouth is full of sharp, pointed teeth, which allow it to hang on to prey securely as it chews. It feeds on a wide variety of aquatic and terrestrial invertebrates, which occasionally include small fish, newts and even frogs. Water shrews are Britain's only venomous mammals, and the mild toxin secreted in the saliva helps to stun larger prey to stop it struggling. If humans are bitten, a tingling sensation is felt and a

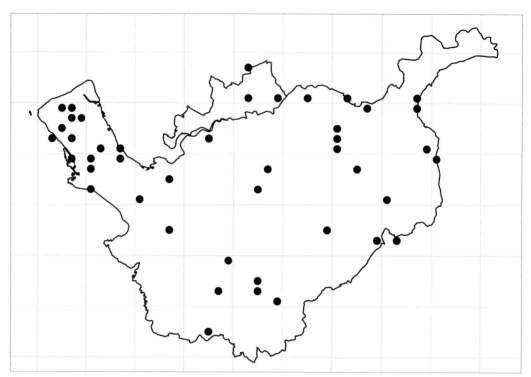

Water shrew pre-2000.

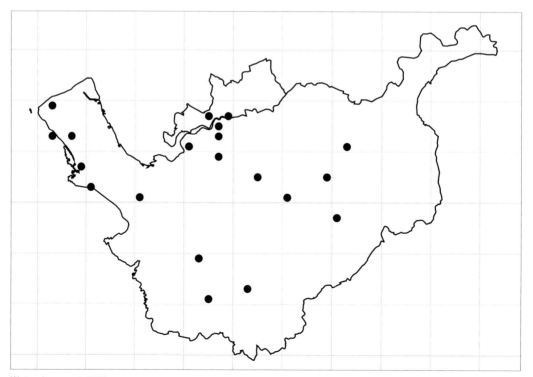

Water shrew post-2000.

red rash may appear around the wound although the bite is harmless and there are no lasting effects. Prey is eaten ashore and the shrew eats roughly its own body weight in food each day, sometimes hoarding food when there is a plentiful supply. Feeding remains can be found at habitual feeding sites.

Water shrews don't hibernate, as is the case with all shrews found in Britain. They are short lived with a maximum life expectancy of 19 months and suffer from high juvenile and adult mortality. The breeding season runs from April to September and individuals are sexually mature in the spring following their birth. Litter size averages six young, with two litters per year. Otherwise water shrews live a largely solitary life.

Key identification features

Water shrews are the largest British shrew weighing 8–23 g and are up to 170 mm in length. They can be distinguished from other shrews by their size and their dark brown-black coat and silvery white belly.

Occasional all-black specimens have been recorded. They are well adapted to their aquatic lifestyle with a waterproof coat, and they have a fringe of white hairs on their feet and a double row of hairs on their tails. They have hidden ears, which are only visible as white tufts; these are closed when swimming.

Observation

Due to their highly elusive nature, and therefore our general lack of knowledge of this particular shrew, a national survey was instigated by the Mammal Society in 2004 to try to find out more about the animal's behaviour, distribution and general lifestyle. The results of the survey led to the production of the first water shrew conservation handbook in 2006.

Normally the shy, secretive water shrew is rarely seen, and it is difficult to spot. When walking quietly along the bank of a slow-moving, clear, shallow stream, careful observation may reveal its tiny footprints in the mud, and its shrill squeaks may be heard amongst the vegetation.

7. Bats Order Chiroptera

In 1986 there were just six species of bat known to be breeding in Cheshire. Since then a further four species have been added to the list and others might arrive (or already be here) following changes in distribution that have been observed over the last few years.

A general introduction to the natural history of bats is given below. All British bats are insectivores.

Yearly cycle

British bats follow an annual cycle of hibernation and activity. As temperatures drop in November they move to a site that provides a constant low temperature which does not fall below freezing, together with a high humidity. All myotid species, horseshoes and long-eared bats can be found hibernating in caves, whilst nyctalus species, pipistrelles and serotines are almost never found in caves and seem to prefer trees, piles of rocks or the cavity walls of old (usually abandoned) houses (Waters & Warren, 2003).

Bats regularly wake up during hibernation. Sometimes this arousal may last for a few hours and the bat will settle back into hibernation at the same site. In other cases, if conditions are right the bat may feed and drink, or relocate to another site if environmental conditions have altered.

As spring approaches and outside temperatures rise, bats become more active. They may stay at the hibernation site, foraging when temperatures are warm enough, or may relocate to a temporary 'gathering' roost.

By mid-May most bats will have moved to summer 'maternity' roost sites. These were originally cave or tree sites but many species, especially common and soprano pipistrelles, now find buildings just as suitable. Most UK species are crevice dwellers and will crawl into small nooks and crannies such as under tiles, in cavity walls, or between double glazing. Long-eared and horseshoe bats can be found in attic spaces and may be visible there during the day. Tree holes provide natural roost sites frequently used by species such as noctules and whiskered bats. Most births occur in late June/July. In August, once young are weaned, colonies will disperse to mating roost sites (often trees), and pre-hibernation roost sites, where time is spent foraging to gain weight prior to returning to traditional hibernation sites.

Reproduction

Mating takes place in autumn and the sperm or zygotes (depending upon species) are held in stasis over winter. Fertilisation or implantation occurs in spring stimulated by the rise in external temperatures. Females give birth in the maternity roosts to one or two young, known as pups. In most UK species a single offspring is the norm but twins have been recorded for noctules on mainland Europe. When the pups are born they are nearly one third of the adult size and are left in the roosts whilst the mother goes off to forage. If a roost is disturbed the mother may move the pup. Depending on species, the young are ready to fly after three to five weeks and, once weaned, mothers will leave them to fend for themselves. The highest rate of mortality occurs during this period. This slow reproduction rate and the use of torpor and hibernation mean that, unlike many other small mammals, bats are long-lived with brown long-eared bats having a life span of up to 30 years. However, this also makes them vulnerable to local extinctions if the young are lost over several years.

Daily cycle

Throughout the active season UK bats follow a daily cycle of torpor and arousal, which is dependent upon the external temperature. In early spring on warm days bats become active at dusk and may venture out to feed, but will return to the roost as the night becomes cooler. As spring progresses, nights become warmer and the level of activity increases. By late spring bats can remain active throughout the night, feeding and resting for short spells before returning to the roost at dawn. On warm days in summer they become active before dusk. In autumn as nights become cooler the periods of torpor become longer and eventually bats enter the deep torpor of hibernation.

Bat calls

Bats use a system of calls called echolocation. These are high-pitched signals, often beyond the range of human hearing, which act as a simple radar system. A bat calls as it travels through the landscape and listens for echoes from objects in its path. Bats also use echolocation to hunt, increasing the number of calls as they home in on insect prey to produce what is known as a feeding buzz.

A common way to listen to bat calls is to use a heterodyne detector, a handheld electronic device which converts the ultrasonic signal of the bat into a sound that is audible to the human ear.

Threats to bats

All British bat species underwent a decline in numbers during the twentieth century. These drops in population size can be linked to the dramatic changes in agricultural practices in the early part of the century, leading to intensification of farming methods and increased use of pesticides. This significantly reduced the amount of insects available and also changed the landscape. Hedgerows which provided links between roosting and foraging sites were removed, and dead trees were (and still are) cut down for safety reasons resulting in a loss of roosting sites. Changes in the urban environment such as barn and loft conversions, the replacement of soffit boxes and the use of remedial timber treatments have all reduced the number of available roost sites. The current penchant for hard landscaping in gardens and the removal of hedges because they are labour-intensive has also reduced foraging opportunities for bats. As a result bats have become rare and are therefore protected by law throughout Europe.

Bats in Cheshire

Coward recorded nine bats in Cheshire in 1910: five as common, two apparently rare and two more, the barbastelle and lesser horseshoe where he was not convinced by the evidence for their presence. Currently eight species are known to breed in the county, two more are recorded as present, and there is one historic record for the lesser horseshoe.

There are two unverified records for new species in the county. The serotine was recorded in 1994 around Dunham Massey National Trust property. Currently thought to be one of the less common UK species, the serotine is found mainly south of a line from the Wash to parts of South Wales. However, as the Cheshire record is one of a cluster of occasional records from Greater Manchester and Lancashire according to the Bat Conservation Trust in 2007 and their *Bat Atlas 2000* (Richardson, 2000), it may therefore represent an expansion of the serotine's range. The barbastelle was apparently recorded around Warrington in 2000 but there are no additional records for this species in Cheshire or further north. The current distribution maps in the *Bat Atlas 2000* record the species as present from occasional records in Shropshire. The nearest confirmed record for the barbastelle was at Attingham Country Park, Shropshire, recorded in the early 2000s.

Cheshire also has historical records of the lesser horseshoe bat. In 1910 Coward reported that it had not been seen in Cheshire for 70 years. At that time this species was known to occur in Derbyshire, Yorkshire and Flintshire. On the current mammal database there is one record for this species from Beeston Castle in 1948. Cheshire Bat Group revisited the castle in both 2006 and 2007 to survey for lesser horseshoes but none were found. This species is currently expanding its range with recent records for Shropshire and Staffordshire, whilst there are healthy populations at both summer and winter roost sites in north-east Wales.

Bats can be seen across Cheshire, in the streets of our urban environment, in woodlands, along hedgerows and over most waterbodies (rivers, canals, lakes etc). One of the best places to watch bats is over Budworth Mere, where up to seven species gather on summer evenings to feed over the water and through the woodland that makes up Marbury Country Park. Guided bat walks are organised by the County Bat Group, Wildlife Trust and Ranger Services throughout the summer.

Breeding in Cheshire	Recorded in Cheshire	Unverified reports
Common pipistrelle	(Current)	Serotine (1994)
Soprano pipistrelle	Nathusius' pipistrelle	Barbastelle (2000)
Noctule	Leisler's bat	
Daubenton's bat		
Whiskered bat	(Historic)	
Brandt's bat	Lesser horseshoe	
Natterer's bat		
Brown long-eared bat		

Bats in Cheshire: records for species since 1900.

Whiskered bat *Myotis mystacinus*
Brandt's bat *Myotis brandtii*

Order: *Chiroptera*   Family: *Vespertilionidae*   Genus: *Myotis*

Status and history

The first record of these species was a specimen found asleep on a wall in the Goyt Valley near Fernilee on 30 May 1885. Coward speculated that this was perhaps the most widely distributed bat in the county, being particularly abundant in the wooded valleys of the Dane and Goyt.

Only identified as distinct species in 1970, the differences between whiskered and Brandt's bats unfortunately do not allow identification in the field without specialist knowledge. Most recent records involve bats at roost sites or specimens found dead or injured.

Whiskered bat.

UK and worldwide distribution

They are widespread in Europe and palearctic Asia, though true distribution cannot be ascertained due to confusion between the two species. Nationally the whiskered bat is considered uncommon with a distribution throughout England and Wales, whilst Brandt's bat is thought to be rare with similar distribution, though absent from Ireland.

Description

Whiskered and Brandt's bats are crevice dwellers and in summer are found roosting in trees and buildings, preferring features such as stone walls and slate roofs. In winter Brandt's are found in small numbers in caves and mines, usually deep within a crack or crevice. Whiskered bats can be found in exposed places or wedged in crevices in cave systems either close to the

entrance, or often in the more humid parts of the cave than the Brandt's.

Females generally give birth towards the end of June and the single young is weaned by six weeks. On average the maternity colony ranges in size from 30–200 but in Cheshire only colonies of fewer than 10 animals have been found so far.

Whiskered and Brandt's bats have similar diets feeding mainly on flies (especially craneflies), spiders and occasionally on moths. Both species will occasionally pick prey off foliage. The whiskered bat is thought to be more often associated with rivers than woodland edges, favouring narrow rivers with dense bank-side vegetation and only occasionally using woodland rides or open areas. In contrast Brandt's bat avoids open aspects, foraging in both broad-leaved and coniferous woodland, forest edges and clear felled areas.

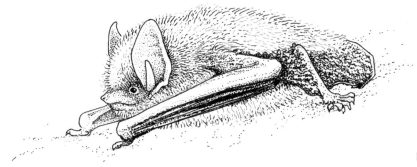

Brandt's bat.

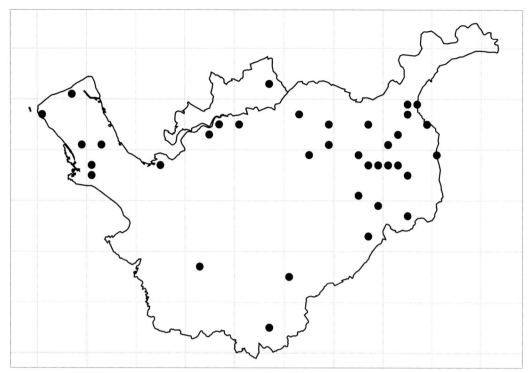

Whiskered bat pre-2000.

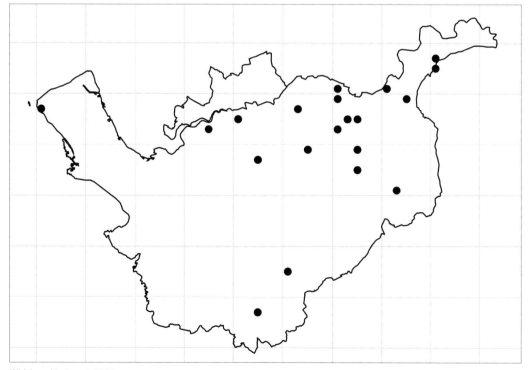

Whiskered bat post-2000.

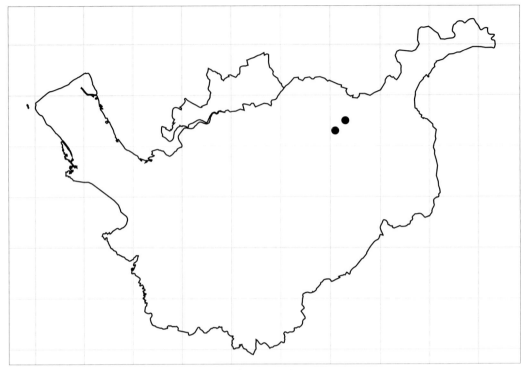

Brandt's bat post-2000. There are no pre-2000 records for this species.

Key identification features

Both species are small and similar in size to the pipis-trelle bat. The whiskered bat has shaggy, dark brown fur on the back and light grey on the ventral side, whilst Brandt's bat is generally golden brown on the back and light grey with a yellowish tinge on the underside. The teeth can be used to identify the two species. Males are easier to tell apart in the hand as the Brandt's has a club-shaped penis whilst the whiskered bat has a thin tapered penis.

Whiskered bats emerge 30 minutes after sunset and fly along habitat edges at head height, fast and straight, with the occasional dive. They glide briefly, especially when feeding in the canopy. Brandt's bats also emerge 30 minutes after sunset and fly fast and straight at head height along habitat edges. In woodland their flight appears more skilful than that of a whiskered bat. On a heterodyne bat detector both species have a call with peak intensity around 50 kHz and can easily be confused with the common and soprano pipistrelle.

Observation

These species are rarely recorded in Cheshire and are possibly underrecorded. One or both species may be observed over Budworth Mere at Marbury Country Park. There is only one known whiskered bat tree roost site in Cheshire.

Natterer's bat *Myotis nattereri*

Order: *Chiroptera*   Family: *Vespertilionidae*   Genus: *Myotis*

TOM McOWAT

Status and history

Coward thought this species to be rare in Cheshire and had only one record: a stuffed specimen, *circa* 1868, which was thought to have been killed near Congleton. At the turn of the nineteenth century Natterer's bat appears to have been underrecorded in Cheshire and neighbouring counties; the first modern record is from 1958 in Tarporley. Although the number of Cheshire records implies that this species is uncommon this probably reflects the trend for recording roosts and not bats in flight. On most surveys or walks near water Natterer's bat will be present.

UK and worldwide distribution

Found throughout much of Europe into Asia with populations on the Pacific coast, in Britain it is present everywhere apart from the north of Scotland and the Western Isles; it is also widespread in Ireland.

The 2007 mammal update reports that the Natterer's bat population has been increasing in the UK since the mid-1990s.

Description

The species is mainly found in woodland habitats ranging from large gardens and parkland to dense woodland including coniferous forests. Trees and buildings are the preferred roost sites, holding up to 200 bats in nursery roosts, up to 25 per cent of which may be male; they will also use bat boxes. Natterer's are frequently associated with old stone or timber-framed buildings such as barns and churches, where they will roost within mortice joints and stonework. After mating in the autumn, maternity colonies are formed from May, with young born from the end of June. Roost sites are changed frequently.

Hibernation takes place almost exclusively in caves or mines. Like many species it swarms in late summer and autumn, usually around cave and mine entrances. During hibernation they are often found tightly squeezed into cracks and crevices but if the site is undisturbed they may hang in the open.

The Natterer's is a gleaning bat taking flies, beetles, moths, bugs and spiders from foliage. Given the choice they will take relatively large prey which may be eaten in flight or taken to a feeding perch; they will often catch prey using the tail membrane. There is a lull in activity before the bats return to the roost.

Key identification features

A medium-sized bat, Natterer's bat has a wingspan of 25–30 cm and weighs 6–12 g. Dorsally the long and shaggy fur is light brown, and buff-white beneath. The ears are large and slightly turned back at the tip; the tragus is long and pointed. A characteristic feature of this bat is a fringe of stiff bristles along the trailing edge of the tail membrane. The echolocation calls of Natterer's bat are very quiet, with a frequency range of 20 to 101 kHz and a peak at about 50 kHz. On a bat detector the calls are heard as irregular clicks and are often described as sounding like burning stubble.

Natterer's bats emerge 40–60 minutes after sunset (through if the roost is within a woodland they might emerge earlier) and are fast and agile in flight, flying at head height along habitat edges and over water at the sides of large waterbodies. They are also able to hover for short periods and, when flying over water, are usually foraging slightly higher than Daubenton's bats.

Observation

Recently Natterer's bats have been recorded at Beeston Castle.

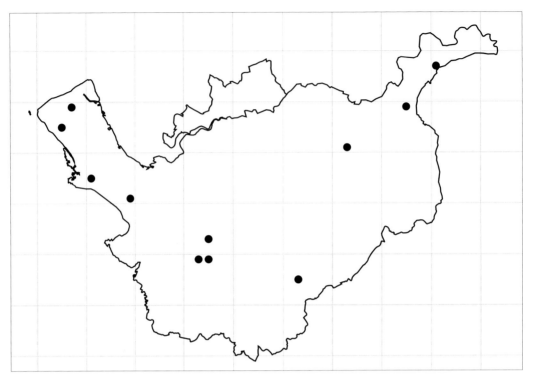

Natterer's bat pre-2000.

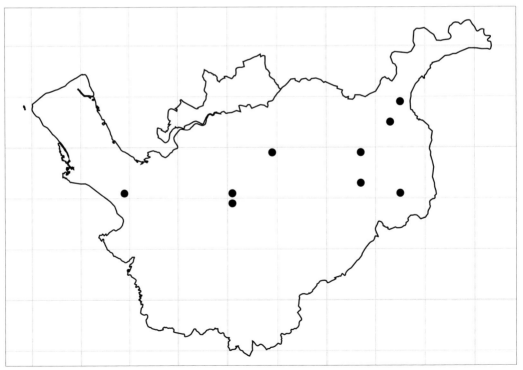

Natterer's bat post-2000.

Daubenton's bat *Myotis daubentonii*

Order: *Chiroptera* Family: *Vespertilionidae* Genus: *Myotis*

Status and history

Also known as the water bat from its habit of skimming the water surface for insects, Daubenton's bat is one of the most widespread of British bats. In 1854 Isaac Byerley mentions nine bats taken by Nicholas Cooke from a hollow tree in Delamere Forest and identified as this species at the Natural History Museum. However, there is some confusion around this record as the specific name *daubentonii* was actually used for the barbastelle, and there is no mention of Daubenton's bat in Dobson's *Catalogue of Chiroptera* published after Cooke's discovery. Unfortunately neither the specimens nor records of their occurrence have been traced.

In 1892 Daubenton's bat was recorded hibernating in the old copper mines at Alderley Edge. Since then the species has been observed in many places, frequenting large waterbodies as well as smaller ponds, moats and slow-flowing reaches of rivers. It was also observed at the east of the county at an altitude of 300 m. Most recent records are from a variety of sources, including the National Bat Monitoring Programme which involves regular surveys of a number of waterways in the county. Other records come from casual site visits or organised guided walks. Few roost sites are known.

UK and worldwide distribution

This is one of the common and most widespread bats in England and Wales but more scattered in Scotland and absent from the Isle of Man and some of the Scottish islands. It is also widespread in Europe and palearctic Asia.

Description

Roost sites are rarely far from water, often in tree holes or in the stonework of bridges; buildings are also occasionally used. One roost may be used for weeks on end or roost sites may be switched on a regular basis. Maternity colonies are established during late spring with the young born from June to early July. They can contain from around 20 to several hundred bats, with up to 25 per cent of the bats being males, including mature adults. Swarming takes place during August and September and mating probably occurs during this time. Like other myotid species, Daubenton's bats usually hibernate underground in caves, mines or suitable tunnels, hibernating either in the open or in crevices.

DAVID QUINN

Insect prey are taken both in the air and from the surface of the water, the latter being 'gaffed' by the large feet or caught in the tail membrane. Although Daubenton's will also forage in woodland, the main prey items are aquatic flies, with caddis-flies taken where abundant. Studies have shown that this species forages night after night over the same stretch of river with favoured sites being stretches of smooth water with good tree cover on one or both banks. Daubenton's bats typically forage up to 3 km from the roost site but are quite capable of flying 15 km along a river during a night's activity.

Key identification features

Daubenton's bat is a medium-sized species, weighing 7–15 g with a wingspan of 24–28 cm. Dorsal fur is sleek uniform brown, ventrally the fur is silver grey, and ears are short with a blunt tragus. Characteristic features of this species are the long calcar (a cartilaginous projection from the foot along the edge of the tail membrane) and very large feet. Ultrasound calls range from 30 to 87 kHz with a peak at 48 kHz. On a bat detector, the calls are heard as a series of regular clicks in bursts of 5 to 10 seconds sounding like machine gun fire.

Observation

This species is easily observed due to its habit of feeding low over water, especially ponds, lakes and smooth stretches of rivers, streams and canals. Emergence usually takes place 30–100 minutes after sunset. Budworth Mere and Macclesfield canal are good places to watch Daubenton's bats.

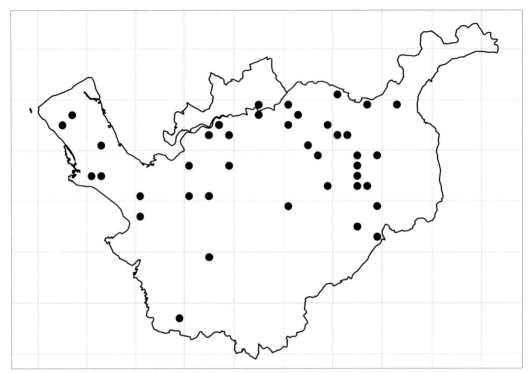

Daubenton's bat pre-2000.

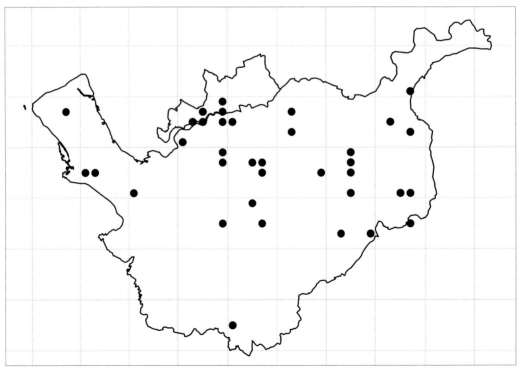

Daubenton's bat post-2000.

Common pipistrelle *Pipistrellus pipistrellus*

Order: *Chiroptera*   Family: *Vespertilionidae*   Genus: *Pipistrellus*

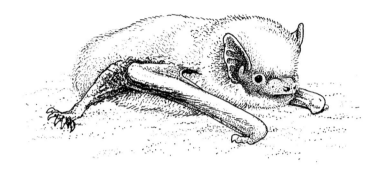

TOM McOWAT

Status and history

In 1910 when Coward wrote the *Fauna of Cheshire*, only one species of pipistrelle was recognised in Britain. Known as *Pipistrellus pipistrellus* this small vesper bat was widespread and common with a distribution from Ireland across continental Europe. In 1997 DNA analysis confirmed that the two phonic types regularly recorded with bat detectors were so different that they were actually separate species. These are now recognised as the common pipistrelle (*Pipistrellus pipistrellus*) which echolocates at 45 kHz and the soprano pipistrelle (*Pipistrellus pygmaeus*) which echolocates at 55 kHz.

UK and worldwide distribution

Coward described the pipistrelle bat as generally distributed across the county and abundant, but he was of course recording the two species together. At the turn of the nineteenth century this was thought to be the most common bat species in the county. Currently there are a number of records of *Pipistrellus* sp. prior to 1997, where species is uncertain. Post-1997 most records are split into the two common species.

Description

The common pipistrelle is still the most commonly recorded bat in Cheshire, probably due to its fondness for modern houses as summer roost sites. Here they are a regular sight feeding over gardens at dusk as well as being occasional accidental visitors inside houses.

The abundance of this species is due to its adaptability: they can be found from upland areas to inner cities and forage in any site where there is sufficient vegetation to support an adequate insect population. Emergence is about 20–30 minutes after sunset (and even occasionally before sunset) when the bats fly directly to foraging sites typically within 2 km of the roost site. Diet is mainly small flies, and up to 3,000 may be consumed in any one night.

Common pipistrelles form maternity colonies in trees, buildings (often new ones) and bat boxes. These colonies of less than 100 bats are highly mobile and relocate frequently throughout the breeding period. Females give birth to a single pup (occasionally twins) in late June/early July. The young are weaned by six weeks and the maternity colonies then start to disperse.

As a crevice dweller it can be found under hanging tiles, behind bargeboards, soffits or eaves, and down cavity walls. Few common pipistrelles are found during winter. They are uncommon in cave and mine systems and usually turn up when a derelict building is being demolished, having secreted themselves in a crevice or cavity within a wall.

Key identification features

One of Britain's smallest bats, the common pipistrelle has a wingspan of 18–25 cm and weighs 3.5–8.5 g. The dorsal fur is nearly black at the base turning brown at the tip. The ears are short and black with a short, curved, blunt tragus. The muzzle is black and slightly longer than the soprano pipistrelle, the bare skin extending back as far back as the ears giving the bat a distinctive mask. Echolocation calls range from 40 to 83 kHz with a peak in intensity at 45 kHz; a social call is also emitted at 20–30 kHz. On a heterodyne detector the call is made up of irregular loud wet slaps which change to clicks at the higher end of the frequency range.

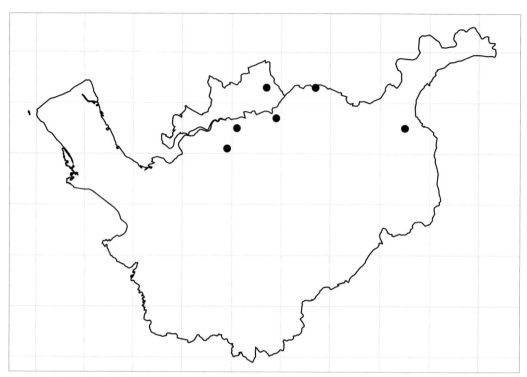

Common pipistrelle pre-2000.

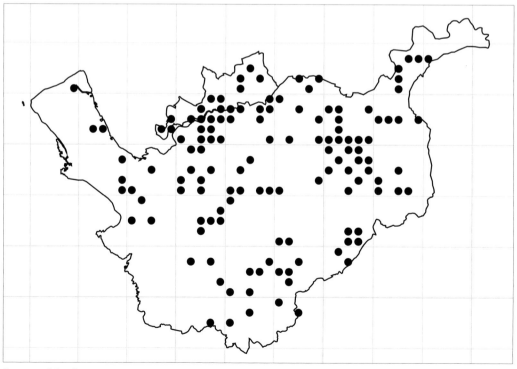

Common pipistrelle post-2000.

Observation

Common pipistrelles hunt along tree and hedge lines, over gardens, along woodland edges and over water, and any other place where small insects gather. Often they will follow a beat flying backwards and forwards, sometimes in a figure-of-eight pattern. Flight is fast and erratic with regular changes of direction just above head height.

Soprano pipistrelle *Pipistrellus pygmaeus*

Order: *Chiroptera*   Family: *Vespertilionidae*   Genus: *Pipistrellus*

Status and history

It is difficult to understand distribution and population trends for this species in the county given the reduction in the number of records following the split of the two pipistrelle species. Currently the distribution of the soprano pipistrelle in Cheshire is similar to that of the common pipistrelle.

UK and worldwide distribution

The soprano pipistrelle is widely distributed throughout Britain and Ireland although there has been a suggestion that the large soprano roosts are more common in Scotland and parts of Ireland.

Description

Soprano pipistrelles form maternity colonies in trees, buildings (old and new) and bat boxes; it is also a crevice dweller and found in the same situations as common pipistrelles. Few soprano pipistrelles are found during winter. They are uncommon in cave and mine systems but have been found in trees, buildings and in crevices in walls and stonework where they appear to be relatively insensitive to cold.

Soprano pipistrelles form large maternity colonies of more than a hundred animals; colonies of over 1,000 individuals have been recorded. These colonies are stable and may remain at one roost site throughout the breeding period. Females give birth to a single pup (occasionally twins) in late June/early July. The young are weaned by six weeks and the maternity colonies then start to disperse.

The soprano pipistrelle has a preference for riparian habitats such as rivers, lakes and meres. It is often found foraging in woodland or trees and tends

DAVID QUINN

to avoid open areas such as farmland and moorland. Like the common pipistrelle it feeds on a wide range of small insects but prefers those with an aquatic larval stage, for example midges, caddis-flies, mosquitoes, mayflies and lacewings. Usually feeding on the wing, they have been recorded gleaning insects from foliage. They often follow the same flight path each night and, like the common pipistrelle, will fly a beat along a habitat feature whilst foraging.

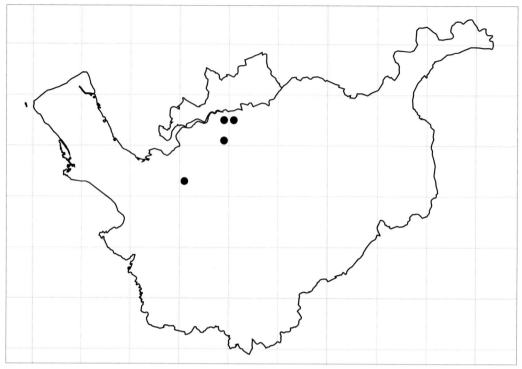

Soprano pipistrelle pre-2000.

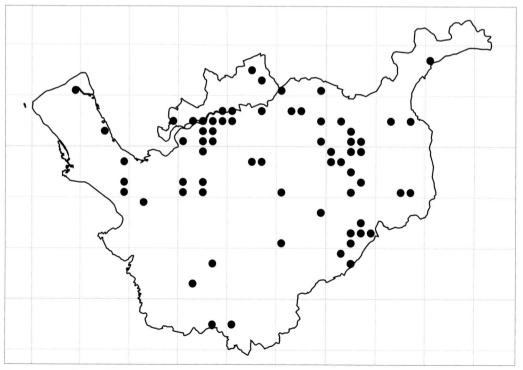

Soprano pipistrelle post-2000.

Key identification features

The soprano pipistrelle is brown in colour, the dorsal fur is dense, smooth and silky. The face and muzzle are light brown, sometimes pink, and more pointed than those of the common pipistrelle, whilst the wing membranes are densely covered with hair.

Soprano pipistrelles emerge 20–30 minutes after sunset and occasionally earlier. The flight is fast and erratic, at head height, along habitat edges. On a heterodyne bat detector the call is made up of irregular loud wet slaps which change to clicks at the higher end of the frequency range. The peak frequency for the soprano pipistrelle is 55 kHz though the call frequency range is 47–90 kHz.

Observation

Soprano pipistrelles have been recorded throughout the summer at Budworth Mere.

Nathusius' pipistrelle *Pipistrellus nathusii*

Order: *Chiroptera*   Family: *Vespertilionidae*   Genus: *Pipistrellus*

TOM McOWAT

Status and history

Nathusius' pipistrelle was first recorded in Cheshire on 23 July 2005 at Budworth Mere, Marbury. The individual was flying over the water at the edge of the mere and foraging. The same species was again recorded at Budworth Mere during the summer of 2006 but no roost site has yet been found.

UK and worldwide distribution

Nathusius' pipistrelle is found from western Europe to Asia Minor. It is rarely recorded in Britain and Ireland, and to date there are only 160 records for this species in the UK. These include four known breeding roosts spreading from Skegness to Belfast. It is classified as very rare with a widespread distribution.

Description

Although considered very rare, the numbers of records for Nathusius' pipistrelle are increasing. In summer, females form large maternity colonies of 50–200 individuals in hollow trees, cracks in trees, bat boxes and occasionally buildings. These roosts are occupied from May onwards with the females giving birth to two young in late July. The males roost singly during this time, probably in trees. Once the young are weaned the females move to join the males in harems of three to 10 females, and mating takes place. Most mating activity occurs in the second part of the night with mating calls being emitted whilst the bat is stationary or in flight. From surveys of hibernation sites on the continent, Nathusius' pipistrelle has been found to overwinter in cliff crevices, wall cracks, caves and hollow trees.

The Nathusius forages by rivers and lakes and in riparian habitats, but has also been recorded in mixed

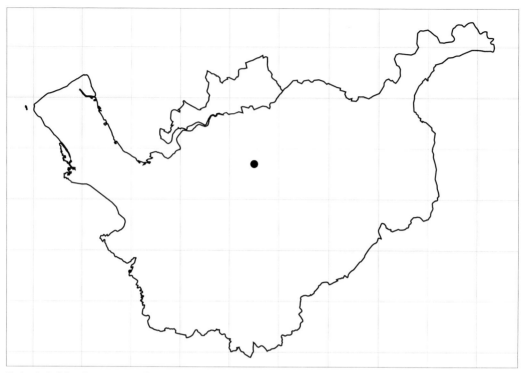

Nathusius' pipistrelle post-2000. There are no pre-2000 records for this species.

deciduous woodland and parkland in the UK. On the continent it has been recorded foraging over agricultural land but unlike the other two species of pipistrelle it tends to avoid built-up urban areas. The main diet is aquatic flies (*Diptera*) including the *Chironimidae* (non-biting midges).

Key identification features

Nathusius' pipistrelle is the largest of the three pipistrelle species found in the UK. The upper fur is longer and shaggier than that of soprano and common pipistrelles, occasionally with pale frosted tips, but generally mid-brown. The underfur is distinctly pale and the face, ears and membranes are generally dark. Ears are medium sized and rounded, slightly longer than broad, with a tragus that has a distinctly blunt, rounded tip. The dorsal surface of the tail membrane is well haired towards the base and beside the tibia. The fur on the underside of the tail membrane sometimes extends obviously along the forearm to the wrist.

The Nathusius usually emerges before the common and soprano pipistrelle at early dusk, 20–30 minutes after sunset. It has a rapid flight, faster than the other two species, and flies from head height to around 4–5 m above the ground, often following distinct habitat edges like hedgerows. Nathusius' pipistrelle is not as manoeuvrable as the common or soprano when flying through vegetation. On a heterodyne bat detector the calls are similar to the other pipistrelles but end at a lower frequency of around 40 kHz. The Nathusius is thought to be the most vocal of the three pipistrelle species, emitting calls upon leaving the roost and continuing all through the night.

Observation

So far this species has only been recorded at one Cheshire site—over Budworth Mere, at Marbury Country Park.

Brown long-eared bat *Plecotus auritus*

Order: *Chiroptera* Family: *Vespertilionidae* Genus: *Plecotus*

TOM McOWAT

Status and history

Coward notes that the brown long-eared bat is widely abundant and known locally as the 'Horned Bat' probably because its ears curl back like ram's horns when the bat is at rest. At the turn of the nineteenth century this species was abundant in Wirral and the Cheshire Plain, and was found at considerable altitude on the hills in the east of the county. During winter months they were found hibernating in the old copper mines on Alderley Edge but moved into attics and cellars for the summer. This is a species which likes the big houses of Cheshire and frequently turns up in roosts in Prestbury, Alderley Edge (where records date back to 1888), Mere and other wealthy areas. It is still recorded across the county in good numbers.

UK and worldwide distribution

Found throughout Europe with patchy distribution, as far east as Japan, in Britain it is found everywhere except the remote Western Isles and Orkney. It is probably the most common species after the common and soprano pipistrelle.

Description

The preferred habitat for a foraging brown long-eared bat is deciduous woodland, parkland or mature gardens within 1–2 km of the roost site. They commute between foraging sites and roosts following linear landscape elements such as hedgerows, tree lines and sheltered rides, rarely venturing out into open areas. They are gleaners, flying slowly among the foliage, and feed by taking moths and other insects such as beetles, earwigs and spiders directly from foliage or other surfaces. They then return to a feeding perch to consume their prey. These perches are found in barns, porches and open structures and are often given away by piles of moth wings and other discarded insect remains accumulating beneath them.

Nursery roosts are found in trees or roof spaces of large, often old, timber buildings, with bats gathering in the ridge roof space, often at the gable end. Long-eared bats spend more time in roof spaces than many other bats, and will fly around inside to warm up flight muscles prior to emergence. Often such sites have significant woodland within 500 m. Roosts generally contain 10–20 bats, although up to a hundred may be present; the sexes are not segregated. Brown long-eared bats show a high degree of roost fidelity. In some cases (such as the bat barn specially built at Manchester Airport) they may remain in the same building all year round, shifting position as the season changes. Known hibernation sites include caves, tunnels, mines and ice-houses although there are no recent reports of hibernation in the Alderley Edge copper mines. The mating period is from October to April and young are born late June to mid-July.

Key identification features

They are medium-sized bats with a wingspan of 23–29 cm and a weight of 6–12 g. Their ears are enormous (two-thirds the length of the body), up to 4 cm in length. The inner margins of the ears meet in the middle of the forehead and the tragus is large and prominent. At rest the ears are folded back and are tucked under the wings during sleep. The fur is long and light brown above, grading to cream underneath; the muzzle is long and usually bare and pink.

Brown long-eared bats usually emerge 45–60 minutes after dusk, though if the roost is within woodland they might emerge earlier. They have a fluttering, butterfly-like flight moving in and out of foliage, occasionally following insects down to the ground; flight often includes steep dives and short glides. They are known as 'whispering bats' because their echolocation calls are so soft. On a heterodyne detector the call has a peak frequency of 39 kHz and sounds like very quiet ticks.

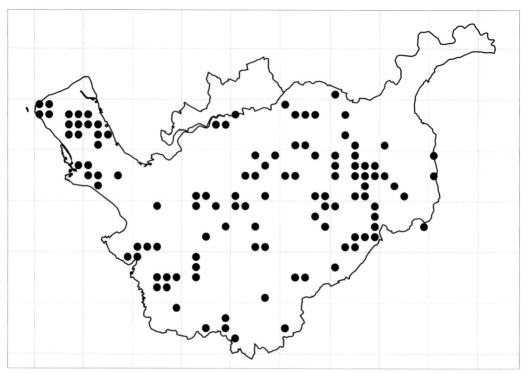

Brown long-eared bat pre-2000.

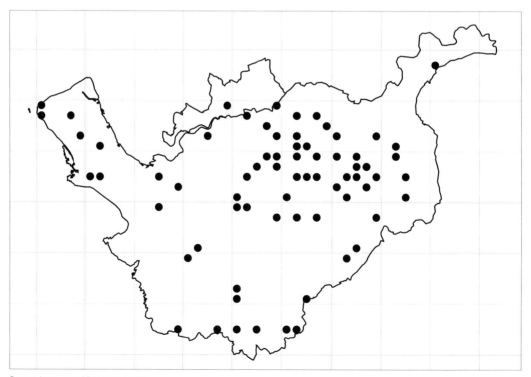

Brown long-eared bat post-2000.

Observation

Flight is rather slow with fluttering wing beats, occasionally hovering, using the large ears to listen for sounds made by moving prey. Tatton Hall, Knutsford is a good place to see brown long-eared bats as there are roosts there, and regular activity over the meres.

Noctule *Nyctalus noctula*

Order: *Chiroptera*   Family: *Vespertilionidae*   Genus: *Nyctalus*

Status and history

The noctule is locally known as the 'Great bat' or the 'Fox bat'. Coward described it as plentiful in lowland areas wherever old timber exists, and abundant in the central plain, especially in large parklands. In the east of the county it was recorded on Alderley Edge and at Broadbottom, both sites at several hundred feet above sea level.

Noctules are widely recorded from sites across the county, either from organised surveys, guided walks or casual records. Very few roosts have been found.

UK and worldwide distribution

The noctule is present throughout England, most of Wales and the extreme south-west of Scotland, but absent from Ireland. It is widespread throughout Europe although distribution around the Mediterranean is patchy, and present across Asia, east to the Pacific.

TOM McOWAT

Description

The Noctule is found in a variety of habitats; as a large, fast-flying bat it regularly travels long distances between roosts and foraging sites and may range up to 26 km from a roost site during the night. It is one of the earliest bats to emerge, and is often seen feeding before dusk. This species usually only forages for 60–90 minutes before returning to the roost, although they may feed again before dawn. The long narrow wings enable rapid flight with frequent swoops and glides. It feeds mainly on flies, beetles and moths taking predominantly large insects, but smaller insects such as midges are taken in spring. Prey is sometimes taken from the ground.

In Britain this species roosts almost exclusively in tree holes but may make use of buildings or other man-made structures and, occasionally, bat boxes. Often roost sites can be identified by urine staining around the entrance. Nursery colonies rarely exceed 20 females but can be much larger in Europe. Males roost singly or in small groups. The commonest hibernation sites are tree holes but again buildings and caves may also be used. Large mixed-sex winter aggregations are sometimes formed.

Males establish a mating roost during late summer which is defended against other males; a series of shrill mating calls is emitted along with a strong odour to attract a harem of four to five females. Mating takes place from August to October and a single pup is born late June to July.

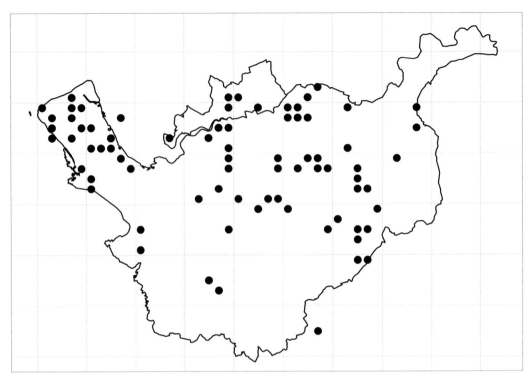

Noctule pre-2000.

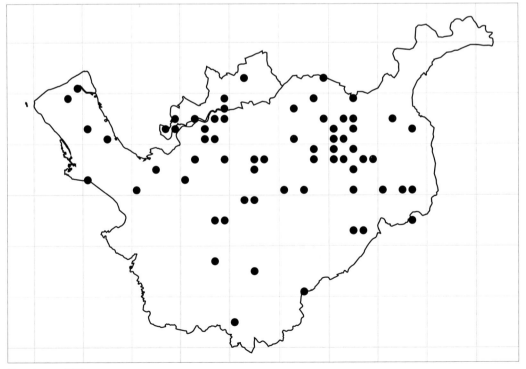

Noctule post-2000.

Key identification features

The noctule is a large, sleek bat weighing 15–49 g with a wingspan of 33–45 cm, bigger than any other British bat. The fur is short, golden or ginger with a well-groomed appearance, the ears are short with a short mushroom-shaped tragus, and the ears and muzzle are dark. Echolocation calls range from 20 to 45 kHz and can be heard by some adults and children; on a bat detector a characteristic 'chip chop' is heard. Noctule calls can be confused with those of the serotine or Leisler's bat.

Noctules emerge around dusk; flight is rapid and takes place in the open, often above trees, but with sudden dives to ground before regaining height. They are also found over pasture, parkland and water.

Observation

An excellent place to see noctules is the reservoir at Trentabank where the evening can be spent watching them fly with the swallows over the water.

Leisler's bat *Nyctalus leisleri*

Order: *Chiroptera*   Family: *Vespertilionidae*   Genus: *Nyctalus*

ROB STRACHAN

Status and history

The first known record for Leisler's bat was a female shot by Coward in Dunham Park in 1899. Another, a male, was shot in 1909 at Broadbottom and in the same year the species was recorded on the banks of the Bollin near Bowden. One of the most recent records for the species in the county was also from Dunham Massey, and further surveys have recorded Leisler's bat in the farmland surrounding Chester Zoo. These recent records suggest the species is still to be found within the county and is possibly being mistaken for the noctule.

UK and worldwide distribution

The third most commonly found bat in Ireland but much scarcer elsewhere, UK records are restricted mainly to central and south-western parts of England with record clusters east of London and in the Derbyshire/South Yorkshire areas. Some of the UK distribution probably reflects migration from Ireland rather than spread of the UK population. The similarity of this species to the noctule has probably resulted in it being underrecorded across its range; it is currently classed as rare in England.

Description

Leisler's bat is less dependent on tree roosts than the noctule and also uses a range of buildings. It has been known to share roosts with noctules and pipistrelles. Leisler's Bat is a particularly mobile species and a roost is often only occupied for a few days before the colony moves on. The average home range is up to 18 km², and foraging areas (including woodland, pasture and riparian habitats) can be 13 km from the roost. Their diet mainly consists of flies, but also

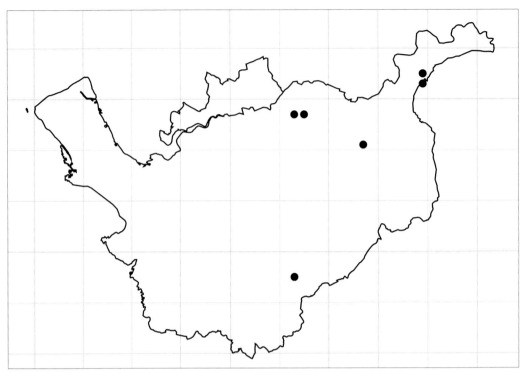

Leisler's bat pre-2000.

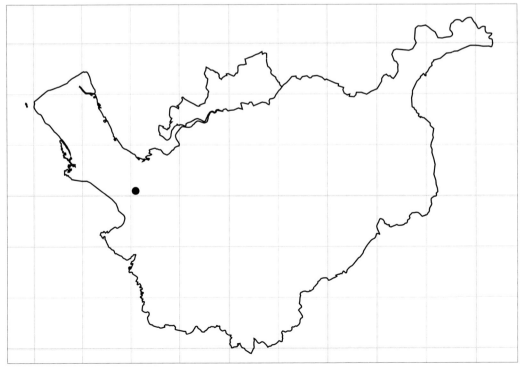

Leisler's bat post-2000.

includes beetles and moths. Foraging usually lasts two hours before returning to the roost, and more foraging trips may be made later in the night. In suburban areas they may be attracted to insects around street lights. Little is known about hibernation biology but, as with the noctule, tree holes seem to be preferred, though occasionally caves and tunnels are used. Maternity roosts are established in late spring, usually averaging 20–50 bats, and young are born mid-June. Breeding occurs from late summer when males emerge from their roosts at dusk and slowly fly around calling. After several minutes they return to the roost to await the arrival of the females. Males may have a harem of up to nine females.

Key identification features

Leisler's bat looks like a smaller version of a noctule, the main difference being that the dorsal fur is bi-coloured for Leisler's bat, with dark roots and red-brown tips. The fur is generally longer than that of the noctule, especially around the shoulders and upper back, giving a lion's mane appearance. Leisler's bat weighs 11–20 g with a wingspan of 26–34 cm.

The Leisler's emerges early, occasionally just before sunset, usually slightly before the common pipistrelle but after the noctule. It has a fast flight, high above the ground from 10 m up to 70 m, in straight lines with fast turns and dives. Over water it will occasionally dip to the surface, or fly lower by bridges. Leisler's bat has a loud call which on the heterodyne detector sounds very similar to that of the Noctule. The call is described as 'chip chip chop' with the peak frequency around 26 kHz within a range of 15–45 kHz.

Observation

Wherever noctules are found then Leisler's bats may also be present.

8. Carnivores Order Carnivora

A carnivore, meaning 'meat eater', is an animal with a diet consisting mainly of meat, whether it comes from animals living or dead. Characteristics commonly associated with carnivores include teeth and claws for capturing and disarticulating prey and their status as a hunter. In truth, these assumptions may be misleading, as some carnivores are scavengers rather than hunters (though most hunting carnivores will scavenge when the opportunity exists).

Eight species of carnivore have been recorded in the region in recent years: six can be described as widespread and still common, the other two are scarce but slowly increasing in numbers. The stoat and weasel remain common but populations may be in slow decline due to fluctuations in populations of prey species. Both species are exceptionally difficult to monitor so accurate population estimates cannot be made. The American mink remains widespread along most watercourses in the county, having a continued detrimental effect on populations of waterfowl and water voles.

Once extinct in the county, the polecat has spread rapidly in the last few years and is now recorded as far east as Macclesfield. An interesting new development has seen more sightings of this species in urban areas where it is now an occasional visitor to gardens.

The badger is widespread and present in good numbers despite still suffering from illegal persecution and high traffic mortality, a common problem for many of our carnivore species. The most obvious of the carnivores is the fox. Not only as it is highly adaptable enabling it to exploit virtually all habitats, but it is also one of the few species active by day. Like the polecat, the otter has been extinct in Cheshire for a number of years. However, with cleaner rivers and improved habitat management, the species is now returning. The pine marten has long been extinct in Cheshire but there have been several records in the Macclesfield Forest area in recent years, although the origin of these animals is subject to debate.

Observing carnivores

Direct observation of carnivores is very much a hit-and-miss affair; often the best results are obtained if the home of the animal is known, and is accessible, without either the animal being disturbed or the observer being seen. If this is not possible there may be an opportunity to observe species such as the stoat in areas where prey is plentiful, or to use artificial feeding to attract them into gardens.

Most sightings of mustelids such as the polecat, mink and weasel, are often just random and generally very brief, all of which makes such species very difficult to monitor accurately.

Fox   *Vulpes vulpes*

Order: *Carnivora*   Family: *Canidae*   Genus: *Vulpes*

DAVID QUINN

Status and history

The fox is a highly adaptable animal which has no specific habitat requirements, and hence can be found anywhere with adequate food and shelter. As most of its foraging and movements are along habitat edges, it is most common in areas where the habitat is fragmented and diverse. Mainly active from dusk until dawn, foxes can also be seen during daytime.

Fox hunting in Cheshire dates back to 1285 when a Royal Charter granted the Abbot of Chester the right to hunt foxes in all the forests of Cheshire. It was not until the latter part of the eighteenth century that systematic hunting of foxes began with the formation of the Tarporley Hunt Club in 1762, followed in 1763 by the Cheshire Hounds, the first regular pack in the county. Several packs have operated for varying periods since then but today only the Cheshire Foxhounds and Cheshire Forest Foxhounds remain active.

Prior to the formation of organised hunts, churchwardens paid a bounty for dead foxes; in 1673 15 heads of foxes were paid for in Rostherne alone. In upland areas gamekeepers and shepherds killed many foxes to protect breeding grouse and sheep and up to 40 were killed during the winter of 1893/94 on the Longdendale Moors. In 2005 hunting wild mammals with dogs was outlawed.

Urban foxes can thrive under favourable conditions, and are often encouraged into gardens by feeding. Road traffic accidents and mange caused by parasitic mites are the main causes of mortality for such animals.

UK and worldwide distribution

The fox is found throughout Britain and Ireland, with populations seemingly increasing in many areas. Foxes are absent from most of the Scottish Islands and the Scilly and Channel Isles. They are also apparently absent from the Isle of Man although reports suggest that the species may have been illegally introduced.

Foxes are present throughout the Northern Hemisphere, ranging north to the Tamyr Peninsula and south to North Africa, central India, northern Burma, North Vietnam, China and Japan. The fox is native to North America although the picture has been complicated by introductions during the middle of the eighteenth century. The species was introduced to Australia in the 1860s and 1870s.

Description

Foxes are territorial and their territory size will depend on habitat, but this can range from as little as 20 hectares up to 450 hectares. Boundaries are marked with urine and faeces. Normally a family group of a

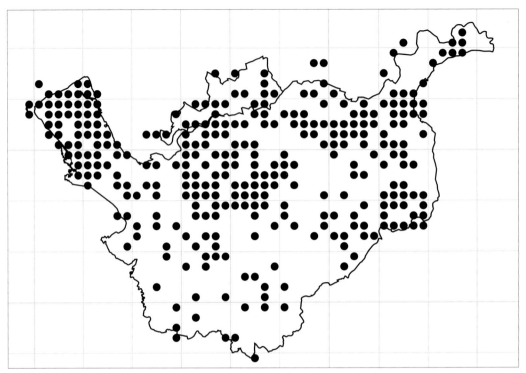

Fox pre-2000.

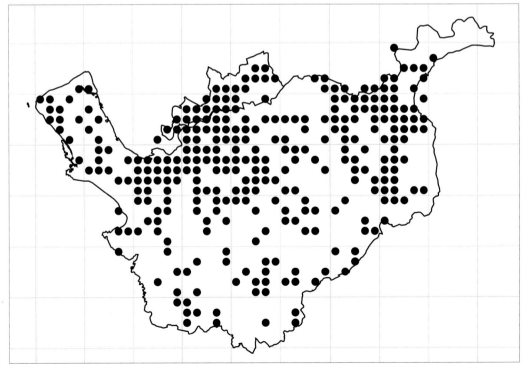

Fox post-2000.

male and one or more females occupy the territory; many females from the previous year's litter may be included. A single litter with five cubs on average is born in spring in a den or earth. This may be dug by the foxes themselves or may be in a disused rabbit burrow or badger sett. In urban areas foxes may give birth under garden sheds. The young remain in the den until they are four weeks old and are weaned by seven to nine weeks. Juvenile animals disperse from the age of six months onwards.

One of the keys to the success of this animal is the lack of specialised food requirements. Most studies show that the preferred prey is rabbits and small mammals such as wood mice and field voles, particularly in lowland rural areas. This may be supplemented by earthworms, beetles, fruit and small birds. Scavenging is also important, especially in upland areas during the winter, with sheep and deer carcasses the food source of choice. Urban foxes will scavenge from dustbins, bird tables and gardens; this diet is often supplemented by food deliberately supplied by householders. If food is abundant, for example, where there are colonies of ground-nesting birds or captive poultry, surplus killing occurs with prey often cached for periods of food shortage.

The fox is mainly a nocturnal animal, but will venture out in daytime and vixens with cubs often play outside in daylight during the summer. The degree of diurnal activity is probably related to persecution. Urban foxes are often most active after midnight although there is much variation. Weather can affect behaviour: warm wet nights are preferred as they bring earthworms to the surface; strong winds and rain tend to keep them under cover.

Key identification features

Coat colour varies from sandy to henna red above with a paler belly. An individual may have a darker stripe along the back. The fore and hind feet are usually black, as are the backs of the ears. The tail may have a conspicuous white tip; this is not, as often believed, confined to dog foxes. Foxes moult in spring, during which time the animal looks thinner and rather scruffy, so can give the impression of being unhealthy though this appearance is quite natural. Head and body length is up to 75 cm with the tail up to 50 cm in length. Males (dogs) are larger than females (vixens) but cannot be separated reliably on size.

Observation

Despite persecution and lack of legal protection foxes are widespread in Cheshire. The population appears to be increasing, particularly in urban areas, although this may be due to greater awareness of the animal. This species is one of the most regularly recorded in the county by both casual observers and via national or local surveys; road casualties in particular provide many records.

Foxes also have distinctive footprints that are ellipsoid in shape with clearly defined claw marks. Fox droppings are easily recognisable and can be distinguished from those of domestic dog due to the presence of prey remains, although old droppings can be mistaken for owl pellets. Feeding signs can also be distinguished: bird carcasses may be consumed very neatly leaving the skeleton with primary feathers attached. Foxes are probably the only mammal that can be readily detected by their odour and, during the breeding season, by their call.

Pine marten *Martes martes*

Order: *Carnivora*   Family: *Mustelidae*   Genus: *Martes*

DAVID QUINN

Status and history

Pine martens used to be found all over Britain and there are records of them throughout the country. By 1850, they were absent or scarce in large areas of southern England, but still present in heavily wooded parts of Sussex, Devon and Cornwall. Probably never common in Cheshire, persecution by gamekeepers and the popularity of the shooting estate, alongside woodland clearance, are all likely to have caused this species to have become extinct in the county some time between 1880 and 1900. The last true Cheshire specimen was shot at Eaton near Chester on 7 July 1891 and the animal was presented to the Grosvenor Museum in Chester, which retains the skin. By 1915, the pine marten was extinct in almost all of Britain, although very small numbers remained in Cumbria, North Wales and the far north-west of Scotland.

Nationally the pine marten population is increasing but only seeming to spread slowly and England and Wales has not seen the scale of recolonisation that might be expected. Reintroductions in Scotland have been successful.

Since 1989 the Vincent Wildlife Trust has received 10 reported sightings from Cheshire, the best being from Rainow in 1992 and Cuddington in 2000. Together with others from Derbyshire and North Staffordshire, these sightings appear to indicate the presence of a population of animals in the Peak District although the possibility of released captive animals cannot be ruled out. Only with increased woodland cover, improved management and better linkages between existing woodland will the pine marten have a chance of becoming established in Cheshire.

UK and worldwide distribution

One of Britain's rarest mammals, the pine marten is chiefly associated with large areas of woodland with old trees offering cavities for the animal to use as breeding sites. The Scottish Highlands remain the stronghold of the animal in Britain. Populations have been reported in the Cheviots, North Pennines, Lakeland, the North York Moors, Peak District and northern and central Wales. It is found throughout Europe apart from the High Arctic, Greece and most of the Iberian Peninsula.

Description

The pine marten is generally a solitary animal with most activity taking place between sunset and sunrise. It is well adapted for climbing having long, well-muscled limbs with strong claws and a long bushy tail to aid balance.

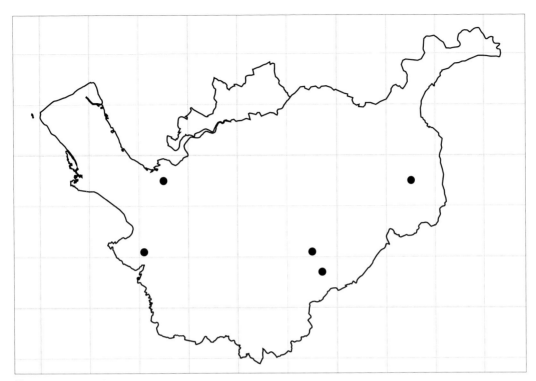

Pine marten pre-2000.

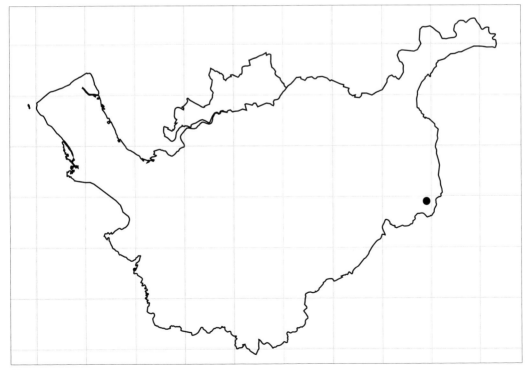

Pine marten post-2000.

Their preferred habitat seems to be closed-canopy woodland, the use of which is influenced by the abundance of favoured prey such as field voles. The animal is equally at home in open, rocky landscapes where crags may substitute for trees and crevices within the rocks provide secure den sites. Dens can be in a variety of places such as hollow or fallen trees, amongst rocks, in old squirrel dreys and even owl nest-boxes; there may be several dens within an animal's home range.

Pine martens breed only once a year with mating taking place in July or August. A delayed implantation (like the badger) sees the pregnancy begin in January. A normal litter of two or three is born March to April. At birth the young have whitish fur and are blind and deaf. They spend at least six weeks in the den before they venture out; the family stays together until they are six months old. After three months they have acquired full juvenile coat and are very active. Individuals may live to 11 years old, but the normal life span is three to four years.

Diet consists of rodents, lagomorphs, carrion, birds, reptiles, amphibians, fish, insects, fruits and berries and varies according to seasonal availability. Pine martens will chase squirrels through the treetops, but these rodents represent a minor part of the animal's diet, so the species provides little threat to red squirrels and would be of little use to control the grey squirrel.

Key identification features

Pine martens are almost cat-like in general appearance and movement but with flatter heads and pointed faces. They may be confused with squirrels when in trees but are much larger and darker. Males may be up to a third larger than females, with a head and body length up to 54 cm and a long fluffy tail of 27 cm.

Upper parts are dark brown with a reddish undercoat. The coat is often darker on limbs and tail. The throat patch is brown and cream or orange and has a unique pattern that can be used to distinguish individual animals. The winter coat is paler and much denser giving the animal a 'chunky' appearance. Ears are large and rounded with pale edges and up to 4 cm in length.

Observation

Pine martens are not strictly nocturnal, especially in summer when most daytime activity occurs. Many animals have learnt to visit picnic sites, bird tables or even kitchen windows to take food deliberately left for them. Local knowledge is especially useful when finding such sites.

Field signs are often difficult to come across owing to the scarcity of the animal and its arboreal lifestyle. The footprint has five toes but the fifth does not always register, thus making it difficult to distinguish from the dog or cat, and the claws may not show unless on soft ground. Faeces are black and cigar-shaped, often containing fur, feathers, bones and other food remains; they are deposited on prominent places within the home range.

Stoat *Mustela erminea*

Order: *Carnivora*   Family: *Mustelidae*   Genus: *Mustela*

DAVID QUINN

Status and history

Writing in 1910 Coward described the stoat as being the most abundant carnivorous mammal in Cheshire. Despite this, records of the animal are few and far between and the national trend appears to show that the stoat population is in decline.

Stoat numbers had been recovering since the outbreak of myxomatosis in 1953 that decimated the rabbit population and thus removed their primary food source. A peak in population size was reached in the mid-1970s (as recorded by the National Game Bag Census) but there has since been a slow reduction despite the fact that rabbit numbers are again on the increase. There remains the possibility that the stoat population will follow suit although there is cause for concern regarding the conservation status of this species.

In areas where rabbits are scarce, the decline in populations of prey species such as water voles, common rats and various farmland birds may be having a negative impact on stoats. Other causes of decline may include increased predation and competition from foxes and potentially buzzards, loss of suitable cover in habitats used for hunting, increasing deaths from traffic and risk of poisoning by taking prey contaminated with insecticides and rodenticides.

The skins of stoats were highly prized by the fur trade, especially the white winter coat, which is referred to as 'ermine' and was a symbol of royalty in Europe. The furs of a number of individual stoats would be sewn together making a pattern of black dots.

UK and worldwide distribution

Present throughout Britain and Ireland and most offshore islands larger than 60 km^2, the stoat is absent from smaller islands due to lack of mammalian prey. It can be found almost everywhere throughout the northern temperate, subarctic and arctic regions; that is, in Europe, Asia, Canada and the United States.

Description

Stoats can be found in any habitat where there is sufficient prey and cover, particularly farmland, woodland, moors and marshes. Open spaces are avoided as the animal prefers to travel along linear features such as stone walls and hedges; it is also scarce in mature forest where little ground cover exists. Sightings are usually fleeting, often of an individual crossing a road or path with the characteristic arched-back gallop or habitually sitting up on hind legs to investigate strange sounds. Stoats are active mainly at night in winter and during the day in summer. They rest in dens distributed throughout their territory, usually constructed in the burrow of their prey.

An opportunist carnivore, a stoat's diet consists mainly of small mammals up to the size of rabbit or water vole, along with common rats, squirrels and small rodents. Prior to myxomatosis, rabbit was the staple food and over 20 years later still accounts for about one third of the diet. It appears that larger prey is selected as this gives the best return in relation to the energy spent catching it. A bite to the back of the

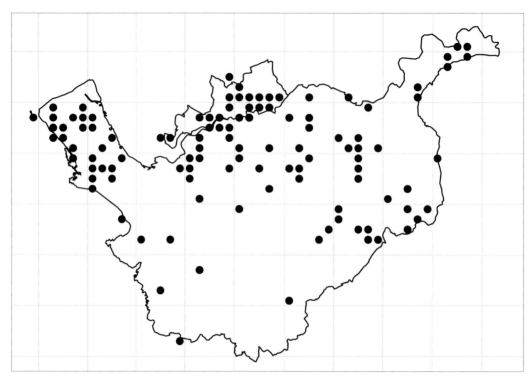

Stoat pre-2000.

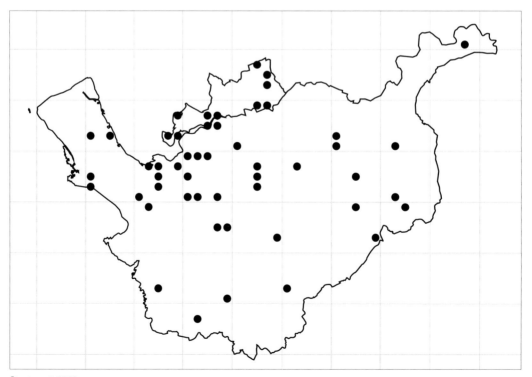

Stoat post-2000.

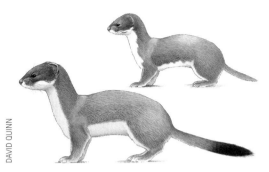

The weasel (top) and stoat compared.

DAVID QUINN

of long dispersal within a few weeks of becoming independent. Young female stoats often leave the nest pregnant, having been fertilised by the resident male. The density of animals is closely related to abundance of favourite prey, particularly in spring.

Key identification features

The stoat is distinguished from its near relative, the weasel, by the distinctive black tail tip that is always present and its larger size with a body length up to 30 cm. In summer, pelage is russet to ginger-brown above and cream-coloured below with the boundary between the two colours defined and straight. In Cheshire, stoats very rarely attain the white ermine coat although intermediate stages may occur; generally the winter coat is denser and paler than in summer.

Observation

Most records of this species are either random sightings of hunting animals, road casualties or those caught by gamekeepers. When observed the first sight is often of the animal standing up, probably to get a better smell of the surroundings. Unlike the weasel this species cannot enter Longworth traps but is occasionally found in traps meant for other species such as mink. Long and twisted droppings, character-istic of mustelids, are often found in prominent places or near the den. However, field signs such as prints and droppings are not reliable as they are similar to those of the weasel and polecat.

neck usually kills, although large rabbits and birds may die of shock. Prey is located by sound or sight; the stoat never lies in wait but actively searches likely habitats with the majority of larger prey animals killed in their burrows. Stoats will take the eggs and young of gamebirds and as a result are trapped by local gamekeepers, although this only appears to cause a temporary reduction in numbers.

As with the weasel, stoats will occasionally fall prey to foxes, dogs and cats but, as with most mustelids, their smelly anal scent glands make them unappetising so they are seldom eaten. Birds of prey will attack stoats but often the animal will use its black-tipped tail to distract the attacker.

A single litter of up to 12 is born in the spring. The young are blind, deaf and hairless at birth and are cared for solely by the female. Fully developed by 10 weeks old, young females usually remain near their natal area whereas the males are capable

Weasel *Mustela nivalis*

Order: *Carnivora*   Family: *Mustelidae*   Genus: *Mustela*

DAVID QUINN

Status and history

Coward considered the weasel to be less common than the stoat. This may have been based on the fact that gamekeepers trapped fewer animals and that they were seen less often during daytime.

Current population trends are unclear but the species does appear to be declining nationally in the long term, the extent of and reasons for this being largely unknown. The 1950s outbreak of myxomatosis reduced rabbit numbers but led to an increase in the small mammal population due to the greater availability of ground cover which in turn benefited the weasel. The subsequent recovery of rabbits and stoats then impacted on weasel numbers. Habitat loss, management practices and the fragmentation and loss of traditional farm features are likely to remain the main threats to weasel numbers and distribution. Trapping and poisoning may be less widely practiced, particularly as reared game is now more common and gamekeepers perceive weasels as less of a threat. Losses caused by road traffic accidents are, if anything, likely to increase in line with the continuing growth in traffic levels.

UK and worldwide distribution

The weasel is widespread throughout mainland Britain and large islands around the UK, but absent from Ireland. It is also found in Europe, North Asia and North Africa. There are two recognised subspecies of weasel: the common weasel (*Mustela nivalis vulgaris*) and the least weasel (*Mustela nivalis nivalis*), but only the former is found in Britain.

Description

Found in a variety of habitats including lowland pasture, urban areas, woodland and marshes, weasels are less common in dense woodland with sparse ground cover as there is little suitable prey available. They are active at all times of day and throughout the year, and will hunt under lying snow in winter. Weasels can take over the nests of prey species and will use several within each individual territory. In cold climates the nests are often lined with fur from prey and they may contain the remains of food from several days' meals. Weasels can occasionally be found under corrugated iron sheets left in grasslands or underneath collapsed stone walls. A weasel's home range varies in size according to the distribution and density of prey; male and females live in separate territories with the male ranges being larger. Resident animals of both sexes may defend exclusive territories at times when numbers are high and neighbours are numerous. In spring males extend their range to seek mates.

Weasels specialise in preying on small rodents; additional prey includes birds and their eggs. Young rabbits are also taken, usually by the males, along with occasional rats and water voles. In general males will take larger prey than females, although this may relate to their different hunting strategies: the smaller female spends longer in rodent burrows so is less likely to encounter rabbits out in the open.

The weasel has a hunting style similar to that of stoats, but spends more time searching the burrows and runways of rodents that are too small for stoats to enter. The head of the weasel is the widest part of

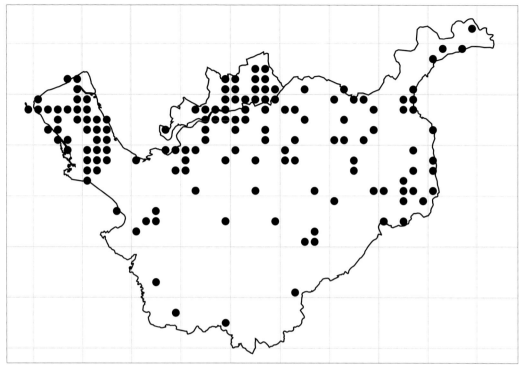

Weasel pre-2000.

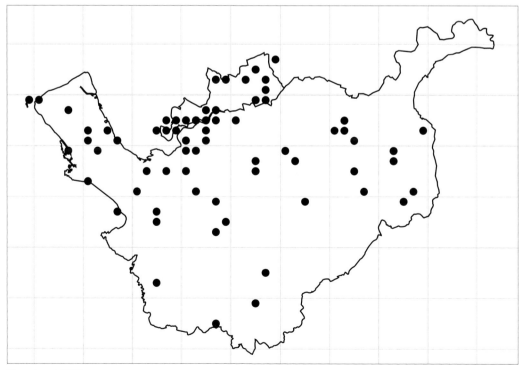

Weasel post-2000.

the body helping the animal to assess where it can go, and the female is able to move through burrows as narrow as 20 mm. Rodents are located by sight, smell or sound and are killed with a bite to the back of the neck.

Probably only one litter of four to six young is produced each season, usually from March to August. They are weaned by three to four weeks and can catch their own prey at six to eight weeks old. Maturity occurs at three to four months, but most do not produce young until the following year, although if voles are abundant the earliest litter may be able to breed.

Weasels are small enough to be taken as prey by almost all other predators such as foxes, hawks, owls, cats and even stoats. Only one in 80–90 weasels survives to two years old.

Key identification features

The weasel has a long slender body and a short tail; the narrow head is supported on a long neck and the legs are short. The large eyes are black and the ears are rounded. Females have a total body and tail length of 20–23 cm and males 23–27 cm: this small size distinguishes the species from all other carnivores except the stoat. Confusion may still arise particularly if the animal is seen at a distance. However, weasels can easily be identified by the lack of black tip to the tail; also the margin between the upper ginger-brown fur and the lower cream-coloured fur is irregular as opposed to the straight margin of the stoat.

Observation

Nowadays, most records are of casual sightings of hunting individuals or the casualties of traffic or gamekeepers. Weasels are small enough to enter and be caught in Longworth traps but this cannot be used as a method of survey. Droppings are typically found at the den or on other prominent places; they are long, thin and twisted and usually contain fur or feathers. Footprints, droppings and feeding signs are very similar to those of stoats, albeit slightly smaller, and again cannot be used as a reliable monitoring tool.

Polecat   *Mustela putorius*

Order: *Carnivora*   Family: *Mustelidae*   Genus: *Mustela*

DAVID QUINN

Status and history

Throughout much of history the polecat has had a bad reputation. The name comes from the French poule-chat, meaning 'chicken cat' due to its reputation as a killer of poultry. It was also known as a 'foulmart' owing to the stench it can produce; its relative the pine marten being known as the 'sweetmart'. The foul smell is a secretion produced as a defensive mechanism, usually by frightened or injured animals. As a further indication of the animal's bad image, in Shakespeare's time the term 'polecat' was used to indicate a vagabond or prostitute. As a consequence the polecat has been persecuted since medieval times, often with a bounty on its head.

The late 1800s saw the rise of the great sporting estates and associated gamekeepers, and also widespread persecution of the polecat due to its habit of taking the eggs of ground-nesting birds, particularly gamebirds. By 1915 polecats only survived in any number in a small mountainous area of central Wales. The First World War saved the species as many gamekeepers left to join the forces, and estate management was never again as extensive. While gamekeepers are now more tolerant and many farmers regard the polecat as an ally against rats and rabbits, it still suffers some (now illegal) persecution.

During the second half of the twentieth century the polecat gradually recolonised the rest of Wales, then began to move into the English border counties. There have been reintroductions into Cumbria and parts of Scotland, but we can be reasonably certain that the polecats in Cheshire are natural colonists from the west.

Data held by rECOrd gives a clear picture of the polecat's recolonisation of Cheshire. The earliest record from Stockport in 1897 is also the last for nearly a century. The next sightings are from Shavington in 1983, Nether Alderly in 1988 and Acton Bridge in 1992. These could be dispersing animals, or equally likely domestic ferrets misidentified as polecats. The situation changed in 1993, with 13 sightings reported, and increasing numbers in every year since. While this may represent increased interest and recording effort, the majority of early 1990s sightings are in the western half of the county, indicating a wave of colonisation from Wales.

An appeal for sightings by Cheshire Wildlife Trust in 2006 brought in about a hundred reports, revealing polecats were in all parts of the county, although distribution is patchy in places and total numbers are unknown. More surprisingly they were abundant in some semi-urban areas, notably Wilmslow, Stockport and Macclesfield.

Hybridisation with feral ferrets identified in some parts of the country is seen as a threat to the survival of the species. The progeny are fully fertile, and after several generations it can be difficult to tell a hybrid from a pure polecat. Scientific studies indicate most Cheshire polecats are genetically similar to the pure polecats in central Wales. Examination of road casualties has also shown no signs of hybridisation in Cheshire

UK and worldwide distribution

The polecat occurs across most of central and western Europe, except the northernmost parts of Scandinavia. It is found as far to the east as the Baltic Republics and the Ukraine. Further east it is replaced by the similar-looking steppe polecat (*Mustela eversmanni*).

Description

Polecats are found in a variety of habitats, including woodland, farmland and scrub. They are often found along river and canal banks and in marshes; this may reflect an abundance of prey in these habitats. They will sometimes come into farms, especially in winter, presumably attracted by rats and mice. In Cheshire they are increasingly being seen in domestic gardens: once again these would offer abundant food to an animal prepared to scavenge for its dinner. Polecats rest up in an underground den; they can dig but prefer a hole already excavated and will often use an old rabbit burrow. They are solitary and territorial; the territory of the male is larger and may overlap that of several females.

The polecat feeds on a wide range of foods, including mice, voles, rats, rabbits, birds and their eggs, frogs, reptiles, fish and invertebrates. Polecats will also happily take carrion, which may partly account for the large numbers killed on roads. In lowland Britain rabbits and rats probably make up the largest proportion of their diet. However, modern rat poisons are persistent, and polecats feeding on rats can accumulate toxins. It is not known how many are killed by this secondary poisoning.

Mating takes place from March to July and is the only time the male and female come together. After mating the male takes no further role in rearing the young. Courtship and mating are a noisy and drawn out affair. To mate, the male grasps the female roughly by the neck, staying like this for up to an hour. This rough copulation is thought to induce ovulation.

Pregnancy lasts 42 days, and the young are born blind and almost hairless. Litter size is usually three to seven, but can be up to 12 kits. The young grow rapidly and the eyes open at five weeks; weaning takes place between six and eight weeks old. The young disperse in late summer and autumn and sadly at this time many become road casualties. They are sexually mature the following year, and can live up to five years (up to 14 years in captivity).

Key identification features

The polecat shares the weasel family body shape: long and thin with short legs and the back characteristically humped when moving. It is easily recognisable by the black and white mask across the face. This colouring can vary through the year, but a white patch around the muzzle is always clearly visible, as are the white ear tips. The neck, legs and tail are black, but on the body the long dark guard hairs are usually fluffed up enough to show the buff underfur beneath, giving it a two-tone appearance. Males are larger than

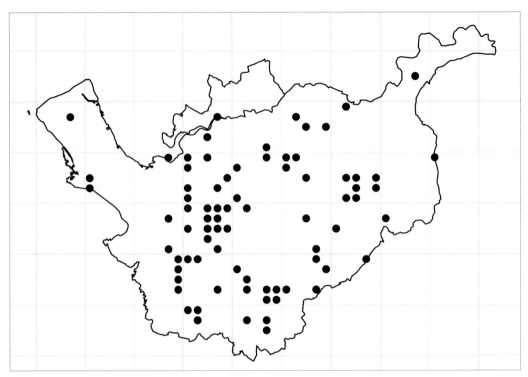

Polecat pre-2000.

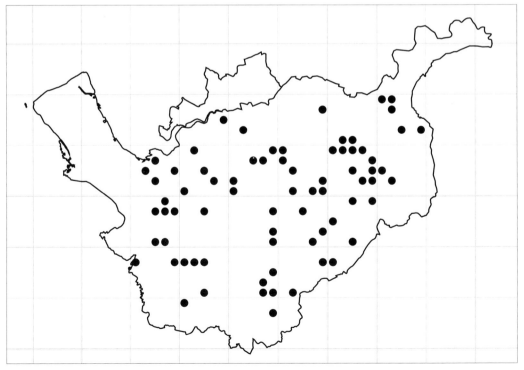

Polecat post-2000.

the females, measuring 45–60 cm from nose to tail tip and weighing up to 1.5 kg; females measure 33–45 cm and weigh up to 800 g.

Observation

Polecats are primarily nocturnal, although they will come out during the day, and the majority of sightings are at dawn or dusk.

The principle field signs are droppings, which are left in prominent places as a means of communicating with other polecats. The droppings are black, up to 7 cm long, often twisted and tapering at the ends, and may contain fur or fragments of bone. They smell strongly when fresh. Care must be taken, as the droppings are similar to those of a mink or large stoat, and indistinguishable from those of ferrets. The tracks are very similar to those of other comparably sized mustelids, and it is impossible to determine the presence of polecat from tracks alone.

Feral ferret *Mustela furo*

Order: *Carnivora*   Family: *Mustelidae*   Genus: *Mustela*

DAVID QUINN

Status and history

It is thought that the Normans brought ferrets to Britain; they were used to catch another of their imports, the rabbit. Currently rECOrd has only two records of ferrets or hybrids in Cheshire, both from 1993, with one in Congleton and one in Burland. The difficulty of distinguishing these animals from polecats may have led to cases of misidentification. However, the examination of several road-kill polecats in recent years has revealed only pure wild polecat characteristics and no traces of ferret. It is concluded from this that there is currently no feral ferret population in Cheshire, and if any animals do turn up they are likely to be recent escapees, therefore no map is presented.

UK and worldwide distribution

The ferret's ancestor, the polecat, occurs across most of central and eastern Europe. Significant populations of feral ferrets occur in New Zealand, where they were introduced in the nineteenth century to control rabbits.

In the UK populations of feral ferrets have established themselves in several areas where native polecats are absent. They seem to be most successful on islands, and there are populations on the Isle of Man, Anglesey and several of the Outer Hebrides.

Description

Until the twentieth century, ferrets were kept mostly for rabbiting. Generally they do not actually catch the rabbits, but are put into the warren to cause the

DAVID QUINN

The polecat (top) and feral ferret compared.

rabbits to bolt into nets placed over the entrances to the warren. More recently, ferrets have also been kept as pets.

When ferrets are placed into a warren it is inevitable that some are lost, due to failure to emerge, or the animals wandering off down the hedgerow. In some parts of Britain these animals have bred to form colonies of feral ferrets. Over several generations the paler animals disappear and the population becomes more like the wild polecat in behaviour and colouration. When they meet, the ferret and the polecat can interbreed, and the hybrid offspring are fully fertile: such hybridisation is believed to be a threat to the continued existence of the polecat. Some sources claim that where they mix, the pure wild polecat easily out-competes the ferret, and that the latter is unlikely to survive, but this is not proven.

Key identification features

The ferret is a domesticated form of the European polecat. It is slightly smaller than its wild relative, and domestication has made it less aggressive. Albinos form a large proportion of the domestic population; the rest come in a wide range of colours, from sandy through to only slightly paler than the wild polecats.

Feral ferrets and ferret-polecat hybrids can be difficult to distinguish from pure-bred polecats. Any animal that has a pale nose, an indistinct mask with more white than black on the face, a white throat or chest patch, white on the paws, or white hairs on the hindquarters is likely to have ferret somewhere in its ancestry.

Observation

As with other carnivores, sightings are often had by chance or specimens discovered dead at the roadside. As previously stated, Cheshire currently does not have a feral ferret population but this does not exclude the possibility of future sightings.

American mink *Mustela vison*

Order: *Carnivora*   Family: *Mustelidae*   Genus: *Mustela*

DAVID QUINN

Status and history

American mink were first imported into Britain in the 1920s to be farmed for their fur. There were many escapes, and some animals were deliberately released. Although there were unconfirmed reports of breeding in the wild as early as the 1940s, it was in the second half of the twentieth century that mink established themselves across lowland Britain.

It seems likely that mink initially spread so rapidly due to lack of competitors: their expansion occurred after otters had disappeared from much of Britain due to pesticide pollution. Now otter numbers are increasing again and, as otters repopulate an area, mink numbers fall. Otters are more efficient fish predators and outcompete mink in the aquatic environment. There is also anecdotal evidence that otters will attack the smaller mink.

The other potential competitor for mink is the polecat, similar in size but with more terrestrial habits. Polecats disappeared from most of Britain due to persecution, but over the last few decades have recolonised most of Wales and the border counties, including Cheshire. There is some evidence that increasing polecat numbers have put further pressure on mink populations. It is unlikely that the return of otters and polecats will push out mink entirely, but will limit their numbers in future.

In Cheshire there are only five records of mink from 1952 to 1976, but during the 1980s the number

of records increases dramatically. While this may partly reflect increased recording effort, it is likely that mink colonised most of Cheshire at this time.

UK and worldwide distribution

Mink are originally from North America, where they are found across most of Canada and the United States. Escapees from fur farming established populations in Britain, Scandinavia and much of northern Europe. In the former Soviet Union thousands were released to establish a new game animal for fur trappers.

American mink have had a profound effect on many other species. In Europe the native European mink (*Mustela lutreola*) has disappeared from much of its range, replaced by its American cousin, a process which is ongoing. It is likely that the European mink will survive long-term only on Baltic island refugia. In Britain, American mink have been the main cause of the massive decline in water vole numbers and have also been implicated in the decline of many riparian birds, although the evidence for this is less clear.

Description

Mink are partly aquatic and are found alongside rivers, lakes and canals, as well as in marshes and in reed-beds; they have also been recorded on estuaries and rocky coasts. Mink need enough cover near water to support

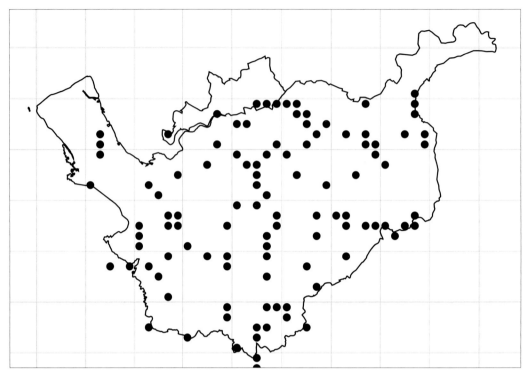

American mink pre-2000.

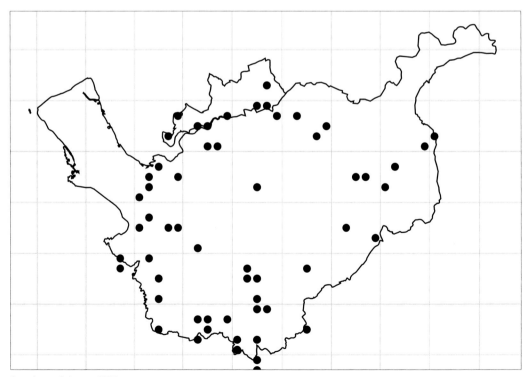

American mink post-2000.

prey populations and provide shelter. Occasionally they are found far from water and can survive for long periods on terrestrial prey such as rabbits and small mammals. These animals are mostly active at night or around dawn and dusk, although they are occasionally seen at other times.

Mink are solitary and territorial. The territories are usually linear as they follow riverbanks or lakeshores and the size of the territory varies with habitat and food availability, from as little as a single kilometre up to six. The territories of a single male may overlap that of several females. Mink mark their territories using scent; glands in the rectum coat the droppings with strong-smelling secretions (known as 'scats'). Within its territory, each animal will have several dens; hollows under waterside trees and old rabbit burrows are often used as den sites. Mink prefer to use existing cavities rather than excavating new sites.

The species are generalist predators, taking a wide variety of foods including mammals, birds, fish and invertebrates. Their pattern of eating varies with time of year and what is most freely available. Rabbits can make up a large proportion of the diet, and smaller mammals such as rats, mice and voles are also taken. Mink will raid the nests of any ground-nesting birds, but water birds, especially moorhens and coots, are their principle avian prey. Coastal mink will also take fledgling seagulls. Mink prefer small, slower swimming fish and eels, and generally hunt in shallow water as, even though they can swim under water, they rarely submerge for more than 10 seconds. They will also take crayfish and crabs. The larger males will often take correspondingly larger prey than females.

A single litter of between four and seven young is born each year; the male plays no part in their rearing. Mating takes place in February and March and at this time males will travel great distances in search of females. Each female is receptive for about three weeks and may mate with several males. Mink have delayed implantation, so pregnancy can vary in length (40–75 days); in Britain most births are in April or May. The young are born blind and hairless. The eyes open at about five weeks, by which time the young are quite active and starting to eat meat. Over the next few weeks they become more adventurous, accompanying their mother on foraging expeditions. They become independent at about 10 weeks old, and soon after this will disperse in search of territories of their own.

Mink are widely hunted, trapped and shot. This makes little difference to their numbers and distribution, and keeping a section of river mink-free requires constant effort.

Key identification features

Mink have the same long body and short legs of all mustelids. Much larger than females, males measure 50–65 cm nose to tail tip and weigh up to 1.8 kg while females measure 48–55 cm and weigh up to 800 g. The coat is generally a glossy dark brown or black although sometimes greys and browns occur, a relic of the many colour variations bred for fur production. There is a small white patch on the chin and throat and the feet are partially webbed to aid swimming.

Observation

Great care must be taken when using field signs to detect mink as the droppings and footprints are virtually indistinguishable from those of polecats, which often use similar habitats. The scats are cylindrical, usually 5–8 cm long, less than 1 cm in diameter and tapered at the ends. When fresh they have a strong, unpleasant smell. They are used to communicate with other mink, so are left in conspicuous places such as on rocks, fallen trees or grassy tussocks. The footprints are sometimes found in wet mud at the water's edge: usually only four of the five toes show in the print and claw marks are often visible at the ends of the toes.

Badger *Meles meles*

Order: *Carnivora*   Family: *Mustelidae*   Genus: *Meles*

DAVID QUINN

Status and history

The badger is still relatively common in Cheshire despite persecution from foxhunters and gamekeepers in the past, and illegal badger baiting in more recent times. The Cheshire and Wirral Badger Group strives to protect badgers and their setts within the county, as well as monitoring the population and raising awareness of the species.

UK and worldwide distribution

The European or Eurasian badger is indigenous to most of Europe and much of Asia, including some parts of China and Japan. It is particularly widespread and abundant (although uneven in distribution) in Britain and Ireland where it has been present for at least 250,000 years. The badger is generally absent from upland regions, large conurbations and areas of intensive agriculture. Badgers are particularly numerous in much of south-west England, and also in parts of the south-east and Wales. Numbers are fewer in most of East Anglia and in the mountainous areas of Wales and Scotland. Estimates of the UK population are about 250,000 to 310,000 badgers living in some 80,000 family groups. The effects of persecution and changing land use mean that they have almost disappeared from some areas.

Description

Badgers are nocturnal and rarely seen during the day. When not active, they usually lie up in an extensive system of underground tunnels and nesting chambers, known as a sett. Occasionally, when the weather is particularly hot, badgers may briefly come above ground during daytime.

They like to build their setts into sloping ground in woodlands, especially where the drainage is good and the soil is not too heavy to dig. Sandy soil seems to be well liked, and heavy clay soils avoided. Hedgerows, scrub, pasture and arable land are also favoured with other important factors being the presence of cover, minimum disturbance by man and other animals in addition to a varied and plentiful food supply.

Badgers are social animals, and a group of badgers living together is called a clan. This is usually made up of several adults and their cubs, and a large clan may have 12–14 adults, plus cubs. Typically, each clan is headed by a dominant male (boar) and female (sow). These animals command the use of the best sleeping chambers in the sett, and the best foraging areas in the clan's range. Normally only one sow in any clan breeds, and mating can take place at any time of the year. After mating, the sow keeps the fertilised eggs in the womb in a state of suspended development until they implant at the end of December; between one

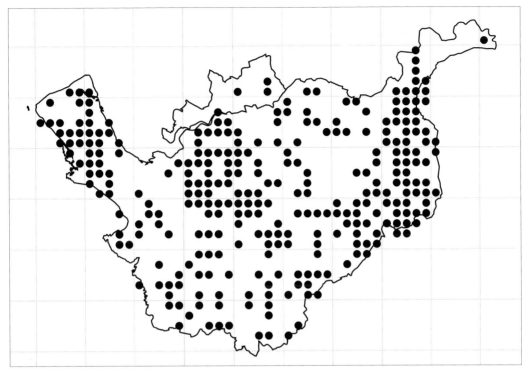

Badger pre-2000.

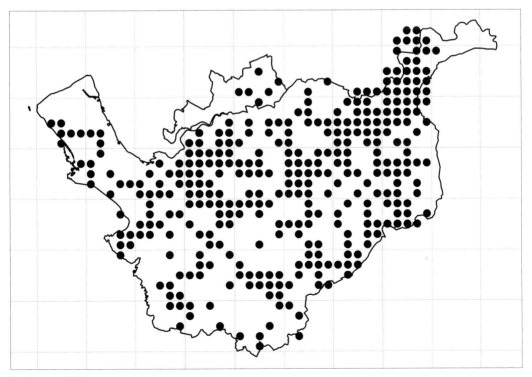

Badger post-2000.

and four young are usually born in February. Cubs are born blind with a pink skin covered in silky, greyish fur. The dark eye stripe is usually visible, and average head and body length is up to 12 cm. Cubs emerge above ground after about eight weeks, and weaning usually starts after 12 weeks, although suckling may continue for up to five months. Dispersal is rare in rural habitats but common in urban habitats, and in both cases usually involves sexually mature males.

Although classified as carnivores, badgers are actually omnivorous and will feed on a variety of animal and plant food according to availability. Earthworms are their most important food item, often caught on pasture or in deciduous woodland, and particularly in wet weather. This can mean that the population will suffer in periods of prolonged drought. Other food items include mammals such as rodents, shrews, moles and hedgehogs; larger insects; cereals; fruit; birds; reptiles and amphibians and carrion. Less activity in winter means badgers are able to survive without food for longer periods, for example, during severe frosts.

Few badgers survive more than six years in the wild; the commonest cause of death is road traffic. Territorial fights between boars may also end in the death of one animal and dogs, foxes and even adult badgers may kill cubs.

Key identification features

The badger is a powerfully built animal with a relatively small head, short neck and low-slung stocky body with short legs and tail. It is the only British mammal with a black and white striped face; the body is grey in appearance and the legs black; its broader head and thicker neck may distinguish the boar. Head and body length is up to 80 cm with a tail up to 19 cm long; males are generally larger than females. A variety of colour variations occur such as albinos, black (melanistic) and rufous/ginger (erythristic) coloured animals.

Observation

Despite nocturnal habits, which make observations of the species difficult, there are many obvious signs of its presence. The sett is large with a number of entrances often with copious amounts of excavated earth and old bedding visible. Badgers deposit their droppings in regularly used latrines and when feeding they leave numerous scrapes in the ground as they search for earthworms.

Their footprints are broader than they are long and despite having five toes on each foot often only four will make an impression. The long claws of the front feet are evident on the better prints. Being creatures of habit, badgers regularly use preferred routes to and from their sett, leaving well-worn tracks up to 20 cm wide. Badger hairs are often left on barbed wire fences and, being long, straight and black and white, they are easy to identify.

Many badgers are unfortunately killed on the roads and, although this provides many species records, it has been claimed that a proportion of such animals have been deliberately killed and dumped along the roadside.

European otter *Lutra lutra*

Order: *Carnivora*   Family: *Mustelidae*   Genus: *Lutra*

DAVID QUINN

Status and history

Once common and widespread throughout the UK, otter numbers declined dramatically from the late 1950s onwards, largely due to pesticide pollution and habitat loss. By 1980 the otter was almost extinct in most of England and parts of Wales and Scotland. With the phasing out of organochlorine pesticides in the UK and the cessation of hunting in the 1960s, the animal has started to make a gradual recovery.

The otter has full legal protection under the Wildlife and Countryside Act 1981 and requires special protection measures under the European Habitats Directive. It is classified by the International Union for the Conservation of Nature (IUCN) as 'vulnerable'.

The otter population in north-west England is being strengthened with colonisation by otters moving south from Scotland into Cumbria and north from Wales into Cheshire. This highly charismatic mammal is once again becoming a feature of our waterways in Cheshire. It is mainly limited to our rivers, streams and canals, but it can also cross watersheds in search of new territory and feed from lakes and ponds. It is the only truly semi-aquatic mammal of the mustelid family.

UK and worldwide distribution

Otters are common in almost all of Scotland and most of Wales. Despite their increase in range, otter numbers are still relatively low in many areas so they remain vulnerable throughout the UK. The species occurs throughout most of Eurasia, to the south of the tundra line, as well as in North Africa.

Description

Nationally otters are now returning to lowland Britain. In Scotland the otter is more commonly seen hunting in coastal waters, requiring access to fresh water to wash and drink. In these coastal areas otter activity is centred around tide times and food availability, but otters in England are largely nocturnal, hunting during the night using their whiskers (vibrissae) to detect prey underwater. Otters rely heavily on fish prey for the majority of their diet, favouring energy-rich, oily meat such as eels although birds, small mammals and amphibians can be important seasonal food sources.

Otters need a wide range of aquatic habitats for resting, breeding and feeding. The loss of wetland habitats in the floodplain, due to factors such as development or agricultural intensification, reduces recolonisation opportunities. Female otters, in particular, require areas of dense cover in order to raise their young. Scrub patches and reed-beds close

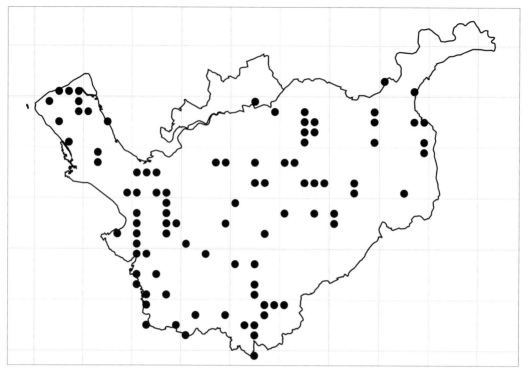

European otter pre-2000.

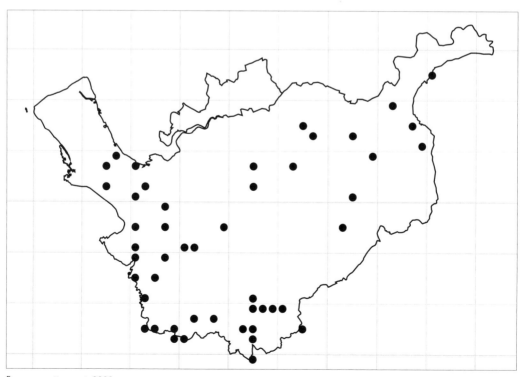

European otter post-2000.

to river systems can provide useful habitats, and the larger and more undisturbed these sites are, the more likely they are to be frequented. Suitable resting sites can include hollows in large riverside tree roots.

One of the reasons for the slow return of the otter is the lengthy breeding cycle. The gestation period is nine weeks and the cubs remain with the mother for up to 12 months. Otters therefore only breed every two years and have between one and four cubs. Life span in the wild is on average four years.

Fiercely territorial, otters are solitary for most of their lives except when breeding and nurturing young. Where food supplies are low, male territories can cover up to 40 km of watercourse although a length of 20 km is more typical. One male territory may overlap with a number of smaller female territories. Droppings known as 'spraint' are used to mark territories.

Key identification features

Otters have a long slender body with medium to dark brown fur that is often pale on the underside. Body length can be up to 120 cm for a large dog otter, of which the tail is approximately one-third and acts as a powerful rudder in the water. Coupled with the webbed feet and streamlined body, this makes the otter a very strong swimmer. Ears are small and flat to the head. Unlike the permanently aquatic mammals, the otter does not have a layer of fat or blubber and therefore relies on a double fur layer to trap air in order to maintain body temperature in the water. The average weights for males and females are 10 kg and 7 kg respectively.

Observation

Road traffic casualties take an increasing toll of otters but are often the first indication that they have returned to an area.

9. Seals Order Pinnipedia

Until the early twentieth century the River Mersey was a lucrative fishery with species such as eel, plaice, dab, sole, shellfish, herring, lobster and mullet. Many of these are prey species for seals and smaller toothed cetaceans. By the 1920s the legacy of pollution from the heavy industries of the Mersey Basin had destroyed the fishing industry but the improvements made in water quality since the 1980s have given a new lease of life to the river. The fish are back and breeding, and so, it seems, are the marine mammals.

Grey and common seals are the most regular visitors. There has also been an occasional sighting of the hooded seal, once in 1873 at Frodsham, when it was captured and exhibited live before being preserved for posterity in Liverpool Museum. The next record is from 1996, when one of these animals was spotted in the north channel of the inner Mersey Estuary, close to the town of Widnes. It remained in the area for some days before it stranded, was rescued and went to a rehabilitation centre. Although this species is usually found in Arctic waters around Greenland and northern Canada, it is known to wander southwards. This one was a particularly intrepid traveller, although there have been sightings off the coast of California!

Grey seal *Halichoerus grypus*

Order: *Pinnipedia* Family: *Phocidae* Genus: *Halichoerus*

DAVID QUINN

Status and history

Coward describes this species as only occurring occasionally in Cheshire waters, probably animals from Anglesey and Caernarvonshire. Prior to this the species had been recorded on three occasions: from Canada Dock, Liverpool in 1860–61, in the Mersey at Warrington in 1908 and one stranded seal at Hoylake in 1909.

This species is now most associated with Hilbre Island in the mouth of the Dee. The first animals were recorded in 1928 when 10 were noted on East Hoyle Bank. There followed a steady increase in numbers, particularly during the 1950s with a peak during 1964 when just over 200 animals were present. Numbers then declined for several years, possibly due to pollution, but have since begun to increase with many recent counts exceeding 500 animals, especially in the summer months.

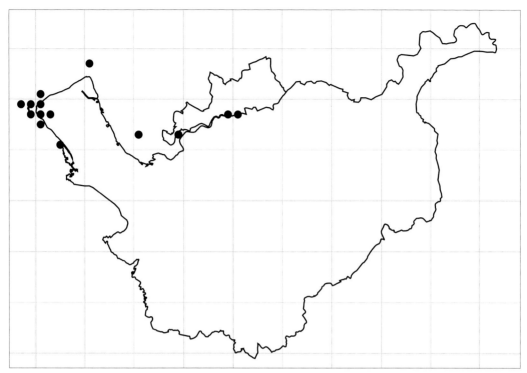

Grey seal pre-2000.

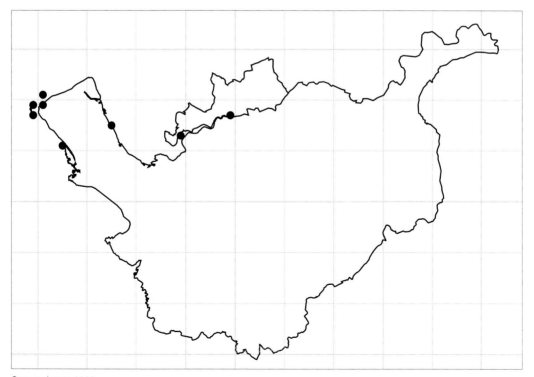

Grey seal post-2000.

It was first thought that the animals at Hilbre originated from the breeding colonies at Ramsey Island off the Welsh coast. Recent radio-tracking studies have since shown that at least a proportion of the population comes from the islands off the west coast of Scotland such as Colonsay.

It is likely that the low numbers of seals present during the early twentieth century were as a result of hunting, hence the absence from Hilbre during this period. There has long been conflict between seals and local fishermen: the animals are often blamed for the decline in fish stocks in and around the Dee Estuary. Investigations carried out during the 1970s with regard to salmon netting off the North Wales coast resulted in no salmon remains being found in gut contents or faecal remains of seals. In recent years, the reduction in sewage entering the Irish Sea appears to have resulted in an improvement in fish stocks as well as an increase in seal numbers.

Sporadic outbreaks of Phocine Distemper Virus have historically caused mass mortalities of seals around the UK coastline and elsewhere in Europe. The last major outbreak in 1988 resulted in the loss of 10 per cent of the UK grey seal population and there are fears that this highly infectious disease will cause further losses in the future.

UK and worldwide distribution

Grey seals inhabit the North Atlantic Ocean with half of the world's grey seal population found on and around British coasts.

Description

Grey seals mainly feed on fish, but will also take cephalopods (squid and octopus) as well as crustaceans.

The species breeds in the autumn. South-western populations begin during September with peak pupping activity late in the month; in Scotland this does not occur until mid-October. They come ashore to breed on exposed rocky shores with the females being the first to arrive at the breeding site or 'rookery' to give birth. When the males come ashore they challenge each other for space nearest to the females. The oldest males get the best positions, but there is minimal fighting. The gestation period is 11½ months, including a three-month delay in the implantation of the fertilised egg. The pup is born with a white coat and weighs about 15 kg, gaining around 2 kg of weight per day thereafter due to the 60 per cent fat content of its mother's milk. After three weeks of suckling the pup, the female will mate again and then leave the rookery. This activity correlates with numbers recorded on Hilbre when the lowest numbers are present during December followed by a rise to a peak in August, a period spent feeding and moulting.

Key identification features

The grey seal is Britain's largest sea mammal. Males may weigh in excess of 300 kg and exceed 2 m in length. Male grey seals are much larger than the females and have broad shoulders, an elongated snout and heavy muzzle with almost parallel nostrils. The females have a thinner snout and a less rounded profile. Colour varies from dark brown to grey or black with blotches, and females tend to be a paler colour than the males.

Observation

Low water is the best time to observe seals on East Hoyle Bank. As the tide flows in the animals soon begin to disperse; most will resume feeding but a number remain in the waters around Hilbre throughout the high tide period. The animals seem very inquisitive, often approaching small boats around the island or observing people on the island itself. There have been sightings of grey seals in the Mersey Estuary in recent years.

Common seal *Phoca vitulina*

Order: *Pinnipedia*   Family: *Phocidae*   Genus: *Phoca*

DAVID QUINN

Status and history

Up to the late nineteenth century the common seal was described as an occasional wanderer to both the Dee and Mersey Estuaries; most records of seals at that time were attributed to this species. There are also a number of inland records including one shot in the River Gowy in 1891 (now in Warrington Museum), a young animal caught in a salmon net and killed in the River Dee at Chester in 1905, one in the Dee at Gayton in 1907 and one killed also in the Dee at Connah's Quay in 1908.

UK and worldwide distribution

The common seal inhabits temperate, subarctic and arctic waters of the north Atlantic and Pacific oceans. In Britain it is widespread along the west coast of Scotland, the Hebrides and Northern Isles with scattered populations occurring along the east coast.

Description

An opportunistic feeder, the common seal feeds mainly on fish, taking species which are locally abundant or easy to catch, as well as cephalopods, gastropods (marine snails and sea slugs) and crustaceans.

Courtship apparently takes place in the water but has rarely been observed. Females give birth to single young in June or July, with the peak in mid-June. Pups are born in intertidal areas, and possibly in the water

itself, with the ability to swim and dive efficiently from birth. Lactation lasts three to four weeks and suckling bouts become progressively longer and less frequent. During this period the pup remains close to the mother who, if danger threatens, will often grab it in her flippers or mouth and dive to safety. After the pups are weaned, the adults mate with gestation lasting 10–11 months including a delayed implantation of two or three months.

Key identification features

In the water the common seal can be difficult to distinguish from the grey seal. However, when hauled out, the common seal adopts a characteristic attitude with both head and tail raised. The head is small in relation to the body, having a concave forehead and dog-like profile in contrast to the grey seal's flat head and elongated muzzle. Unlike the grey seal, the common seal's nostrils meet at the base to form a 'V' shape. The colour pattern is variable, usually dark spots on a lighter background and males are generally darker than females.

Observation

Occasional animals may be seen with the grey seals in the Hilbre area; during late 2006 a young animal was noted in the Mersey off Pickering's Pasture, Widnes.

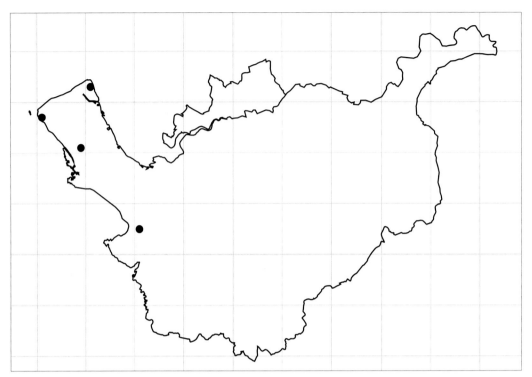

Common seal pre-2000.

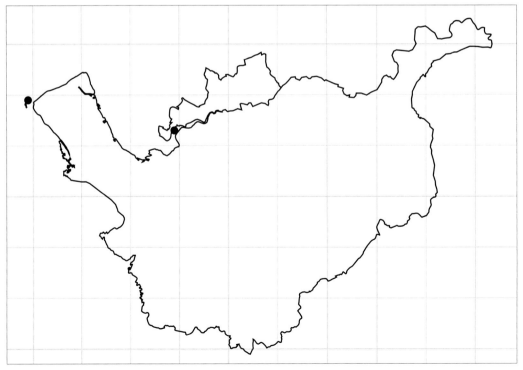

Common seal post-2000.

10. Deer Order Artiodactyla

Deer are Britain's largest wild land-based mammals and there are six free-living species of deer here. Only two of these, the red deer and the roe deer, are truly native to the country and even their populations have, over the years, been heavily 'subsidised' by introductions from elsewhere. The other four are introduced alien species: the fallow deer was almost certainly brought to this country by the Normans and three Asiatic species (Reeves' muntjac, the Chinese water deer and the sika deer) were introduced at the end of the nineteenth and beginning of the twentieth century. Full species accounts are given for the native species and the two most numerous introductions, namely the fallow deer and muntjac, with the other two species (sika and Chinese water deer) described briefly at the end of the chapter.

Deer are from the order *Artiodactyla* meaning 'cloven-hooved', having an even number of functional toes on each foot (usually two). They are ruminant herbivores having a chewing pad instead of upper incisors and a specialised form of digestion with a four-chambered stomach. They have highly specific nutritional requirements and when feeding they tend to select easily digestible shoots, fresh grasses, soft twigs, young leaves, fungi and lichens. Different species have different feeding habits and preferences. Roe and muntjac, for example, are specialist browsers, whereas fallow deer are predominantly grazers. All deer can consume large quantities in a short time and hence have the potential to cause major damage in gardens, plantations and crops.

Active throughout the 24-hour period, deer have peak times of activity at dawn and dusk. Populations experiencing frequent disturbance make more use of open spaces during the hours of darkness; most daylight hours are spent 'lying up' ruminating. Dawn and dusk are therefore the best times to see wild deer, though the managed herds at parks like Tatton, Dunham Massey and Lyme are usually visible at any time. As with all animal watching it is vital to be quiet and patient. Deer have very good senses of hearing, sight and smell, so any approach should be made quietly from downwind without allowing one's silhouette to break the horizon. Field signs such as tracks and droppings should be carefully noted. Deer will usually see you before you see them: most sightings are of fleeing rumps and this is why such features are so important in recognition!

Red deer *Cervus elaphus*

Order: *Artiodactyla*   Family: *Cervidae*   Genus: *Cervus*

DAVID QUINN

Status and history

Thomas Coward declared that red deer were once common in Cheshire forests but by the time of his writing in 1910 they were only present in managed parkland herds. Two of these parks, at Tatton and Lyme, exist today and both still have red deer herds. The Tatton herd may have once contained wild deer from the Cheshire area but there has been the addition of deer from Scotland and elsewhere. The Lyme deer are likely to be, at least partly, descended from wild deer roaming the Macclesfield forest area which were originally enclosed in the fourteenth century.

All recent red deer sightings outside the parks are in the far east of Cheshire. There seem to be two different populations emerging: one from the Goyt Forest/Hoo Moor area and the other from the far south-east in the River Dane/Bosley Minn area. The Goyt Forest population was first recorded in the mid-twentieth century and has spread west into Cheshire from the Derbyshire border. The population from the Bosley area is recorded from the 1970s onwards and the data shows a slight population spread from the south-east of the county in all directions to the mid-1990s. Red deer records are most numerous from this area and the likelihood is that this is a feral herd derived from a captive population at Roaches House near Leek. Animals from this private zoo are known to have escaped during the Second World War.

UK and worldwide distribution

In the UK the species is common in Scotland with scattered populations in Wales and England including the Peak District and the Pennines. Red deer are one of the most widely distributed deer in the world, stretching from Eurasian temperate zones to northern Europe. There are 12 subspecies recognised across this range.

Description

Red deer have a life span up to 15 years in the wild or longer in captivity. They are grazers and browsers feeding on grasses, rushes, heather, leaves and bark.

Within its range in England this animal occurs in woodlands and forests but can adapt to open moor and uplands. In woodlands red deer are largely solitary or occur as mother and calf groups. Elsewhere they live in single-sex herds for most of the year, the female herds including young and non-breeding males. The sexes only meet up in the autumn for the rut (mating) though dominant males will follow female herds from August to early winter. At the rut, stags roar loudly, display and confront each other; fights are frequent in large herds. Roaring most often takes place at dawn and dusk and is audible from some distance. Calves are born in May and June.

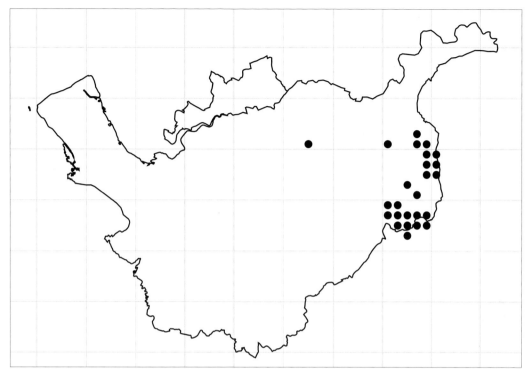

Red deer pre-2000.

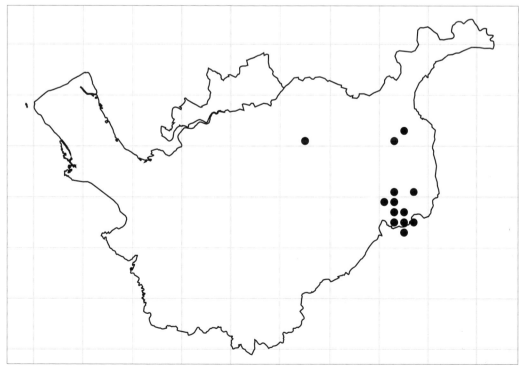

Red deer post-2000.

Key identification features

The red deer is the UK's largest land animal with a shoulder height of 105–120 cm (both sexes) and weight up to 200 kg in adult stags. Hinds are significantly lighter than stags.

During the summer red deer are dark red or brown with lighter cream underbelly, inner thighs and rump. In winter the pelage changes to darker brown or grey, still with cream patches on the rump and undersides. The light-coloured rump patch tends to reach higher up the rump than on other deer. Calves are born brown with white spots which usually disappear by two months old, but adults occasionally retain a few spots.

Both sexes have tails of approximately 15 cm in length, which are generally the same colour as the rump patch although the tail may have a dark dorsal stripe.

Each year males grow spectacular branched antlers up to a metre wide that are then shed in winter. In September and October, prior to the rut, males also develop a muscular neck and a mane.

Observation

In Cheshire red deer are most likely to be seen in woodland or on moors. Field signs include tracks, droppings, evidence of feeding, mud wallows and vegetation trashed by stags with their antlers before and during the rut. in winter, red deer feeding on tree bark leave very obvious evidence and can cause severe damage. Red deer droppings take the form of a mass of dark brown or blackish, oblong-shaped pellets, about 2 cm in length.

Fallow deer *Dama dama*

Order: *Artiodactyla*   Family: *Cervidae*   Genus: *Dama*

DAVID QUINN

Status and history

Introduced by the Normans in the eleventh century and quickly establishing themselves in parks and forests, fallow deer have since become the most widespread species of deer in Britain. No mention of them in Coward (1910) leads us to assume that they were either not present or were very rare in Cheshire at that time. Today there are managed captive herds at Tatton, Dunham Massey and Lyme parks and a significant wild population exists in North Wales in northern parts of Denbighshire and Conwy.

The two earliest fallow deer records on the Cheshire database are from 1971. One record is for Little Eye on the western side of the Wirral, the other is from Liverpool. The recorder is not known for either of these records although the similar dates and relative proximity of the records add weight to their validity. As with most anonymous records they must be treated with caution. The most plausible explanation for fallow deer in Cheshire seems likely to be escaped animals from parks. However these two records perhaps suggest movement into Cheshire from the Welsh hills.

UK and worldwide distribution

Fallow deer are native to the Mediterranean region and from Turkey to Iran. They have also been introduced to 38 countries elsewhere around the world.

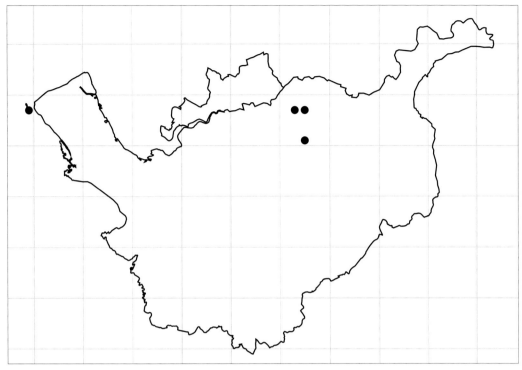

Fallow deer pre-2000.

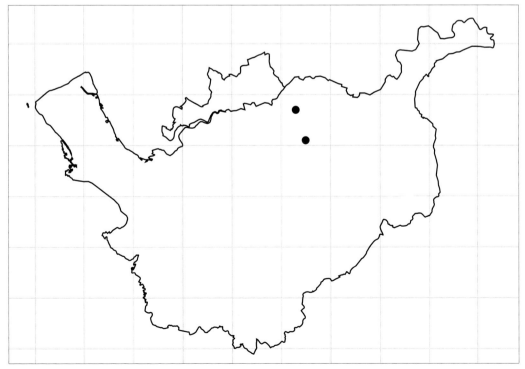

Fallow deer post-2000.

Description

Fallow deer are grazers and browsers feeding on grasses, herbs, berries, acorns, bark, shrubs and the growing shoots of trees; they can cause a great deal of damage feeding on buds and stripping tree bark. They may also feed in arable fields on root crops such as carrots, sugar beet, parsnips or potatoes. Life span is up to 16 years but wild bucks rarely exceed eight to 10 years.

These deer range over large areas spending only a short time in one place and typically occupy deciduous woodland with open patches but will also inhabit open agricultural environments. Group size and the degree of sexual segregation vary according to population density and habitat. In woodland, fallow deer commonly live in small isolated groups, though herds of 70–100 may gather in good feeding areas. Groups commonly consist of adult males or females with young; sexes remain apart for most of the year only coming together to breed. In more open agricultural situations the sexes freely mix in large herds throughout the year.

The rut starts in late September and peaks in mid-October. Competing males display by groaning, thrashing their antlers and by walking alongside opponents. Fighting occurs if two stags are evenly matched; this involves wrestling and clashing of antlers. Does usually produce a single fawn between late May and mid-June leaving the herd to find a hiding place to give birth. After birth the fawn remains hidden in bushes or dense vegetation with the doe returning to feed it until it is about four months old. Fawns then join the mother's herd and are weaned after seven to nine months. Young bucks stay with the doe herds until they are 18 months old when they leave to join the buck herds.

Key identification features.

Fallow deer are intermediate in size between roe and red deer with a shoulder height of 85–100 cm. Weight is 60–85 kg for bucks and 30–50 kg for does. The only British deer with palmate (flattened) antlers, a mature fallow buck has a very prominent Adam's apple and an obvious brush of hair under its belly.

There are several different colour variations of fallow deer in Britain, the most common form being a chestnut coat with white spots on the flanks, and a white rump patch with a horseshoe-shaped black border. This form develops a dull brown winter pelage with spots faint or absent. The paler Menil variety keeps its white spots throughout the year and lacks the black-bordered rump patch. The black variety is almost entirely black with no white coloration anywhere, whilst the white form can be white to sandy coloured, becoming whiter with age. The tail of a fallow deer is quite long with a distinct black line running down it in most colour forms.

Roe deer *Capreolus capreolus*

Order: *Artiodactyla*   Family: *Cervidae*   Genus: *Capreolus*

DAVID QUINN

Status and history

No mention by Thomas Coward in 1910 leads us to assume that roe deer were not present in Cheshire or were very unusual at that time. Very little can be deduced from twentieth-century data as sightings tend to be focused around parkland areas where captive populations of fallow and red deer are present. The earliest record is from Hyde in Manchester by an anonymous recorder in 1918; the grid reference is SJ99 so it is not possible to pinpoint the location. Chronologically the second record is near Warrington in the 1990s with another at Lyme Park in 1994. There are two records for Tatton and Knutsford in 1998 and 1999 respectively. Since 2000 it appears that this species may be becoming more common in Cheshire as there have been more frequent sightings: 2004 at Comberbach, 2005 at Woolston, 2006 (two near Risley, one at Middlewich and one at Marton) and 2007 near Poynton.

UK and worldwide distribution

There is only one species of roe deer, considered to be divided into three subspecies: European, Siberian and Chinese. In the UK, roe deer are common in Scotland with scattered populations in Wales and England including in the Peak District and Pennines. Historically, forest clearance and overhunting led to roe deer becoming extinct in England by 1800 but they remained in wooded patches in Scotland. Reintroduced during Victorian times, roe deer have been aided by twentieth-century woodland planting and are widespread today.

Description

Predominantly selective browsers this animal also grazes, feeding on shoots, herbs, grasses, fruits, nuts, fungi, pine needles and twiggy branches during hard times. They are opportunistic feeders with a taste for exotic plants, as many gardeners have discovered! Roe deer are highly adaptable and occur in a wide variety of habitats ranging from open moor to thick cover in conifer or deciduous woodland. Ideal conditions are coppice and pockets of deciduous woodland on land that is not intensely farmed, with thick hedgerows and scattered copses. They seem to follow linear features such as hedgerows, roads and canals across the landscape.

Bucks are usually solitary, and does will have kids with them most of the year. When a doe appears alone it is likely that there is a buck not far away. Roe deer in captivity can live to 16 years but bucks in the wild rarely exceed five years and does, six to seven years.

Spring is a busy time when bucks establish territories; those without status are pushed to the

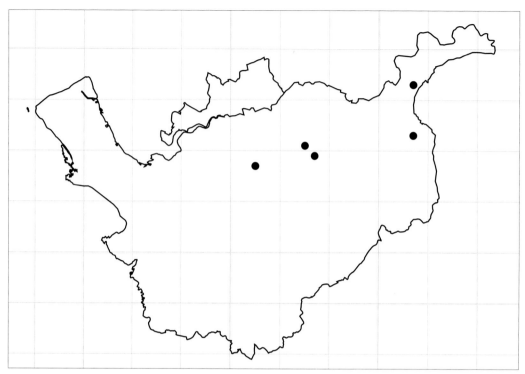

Roe deer pre-2000.

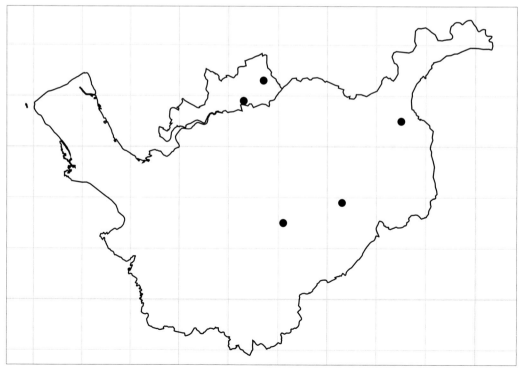

Roe deer post-2000.

periphery of good areas where food and cover are less favourable. Does also drive the previous year's kids from their home range at this time. This leads to emigration and immigration between areas, which is very significant in relation to roe deer distribution. Ousted from familiar surroundings, many animals wander widely seeking a favourable place to live.

The rut takes place between late July and early August. Roe deer are unique amongst British deer in that embryo development is delayed until late December or early January, when rapid foetal development begins. The majority of kids are born in May and June, although they have been seen as early as March and as late as August in the south of England.

Key identification features

Roe deer stand 60–80 cm at the shoulder with bucks weighing 24–30 kg, whilst the does are 2–6 kg lighter. An important distinguishing feature of the species is the very short tail which is not visible without close physical inspection.

Bucks bear short spiky antlers that are shed and regrow in winter. Adults have a characteristic six-point head with minor variations depending on the health of the animal. Young males may have a single spike or forked tine. Buck kids develop a small knob (or button) during the first nine months of life; this is shed in early spring when the first real antlers begin to grow.

At birth, roe kids have a heavily spotted dark brown coat, with two distinct lines of white spots running either side of the dorsal line from the nape of the neck to the rump. The sides and flanks are also spotted in white. The spots soon begin to fade and are usually gone in about eight to 10 weeks when kids develop their first winter coat. By September/October adults are in full winter pelage, coat colour ranging from dark brown to charcoal grey. The caudal patch is white throughout the year but is far more prominent in winter pelage; it is important in distinguishing sexes during the winter when the antlers are not obvious. Does have a distinctive downward pointing tuft of hair at the base of the caudal patch, which gives the appearance of an ace of spades; bucks do not have this tuft and their caudal patch is kidney shaped.

The winter coat is shed from April to May, with younger animals changing first. Whilst moulting, roe deer have an unkempt appearance as their coat falls out in chunks. The summer coat is commonly chestnut red although it can vary to a sandy yellow.

Both bucks and does have a white spot on the upper lip either side of the nose and the chin is also white; these markings are more prominent in young animals. As roe deer age, an increasing amount of grey develops around the muzzle and the hair on the forehead becomes curly.

A short bark, often repeated, may be given by roe deer when alarmed. During the rut the doe makes a high-pitched piping call to attract the buck who makes a rasping noise as he courts her.

Reeves' muntjac   *Muntiacus reevesi*

Order: *Artiodactyla*   Family: *Cervidae*   Genus: *Muntiacus*

DAVID QUINN

Status and history

Introduced to England in the early twentieth century, Reeves' muntjac is now an established feral species, possibly equal to roe deer in its ability to adapt and colonise. Muntjac sightings are focused in central Cheshire. There are three records around Swettenham: two from 2000 by S.J. McWilliam, and the other from 2004 at Stockery Park Farm by McDermott. The earliest recording was in 1969 at Oakmere by an unknown recorder. All other records are post-2000 and seem to show recent movement into Cheshire.

UK and worldwide distribution

Reeves' muntjac are the oldest of all known deer; there are eight species indigenous to South-East Asia. First introduced from China to Woburn Park in the early twentieth century, deliberate releases and escapes of muntjac from Woburn, Northamptonshire and Warwickshire have led to feral populations, the most significant release being from Whipsnade Zoo in 1921. People have assisted their rapid spread across England and Wales although this is now illegal.

Description

Reeves' muntjac have a life span of 10–19 years, though this upper figure is exceptional. They have undiscerning feeding habits and take advantage of diverse food sources. They are predominantly browsers readily eating shoots of shrubs and young trees, ivy, bramble, grasses, herbs, fruit, nuts, berries, fungi, flowers (wild and cultivated) and vegetables. Woodland wildflowers like bluebells, primroses and honeysuckle are particular favourites, and they are unaffected by some poisonous plants including yew and dog's mercury. Muntjac have a unique feeding method for young trees, bending saplings over between their front legs and 'walking up' them to reach high foliage. Feeding normally occurs every three to four hours with bouts lasting 30–40 minutes, after which the animals retreat to cover to ruminate.

Readily colonising any area with an abundance of thick cover and food, muntjac favour woodland but are extremely adaptable. The species is increasingly being seen in urban situations in the south of England and is establishing in areas such as Salisbury Plain which is predominantly open grass and scrubland.

Generally believed to be solitary animals, both bucks and does have their own territories. Bucks are territorial all year round and mark their patch by urinating on raised earth and vegetation, marking boundaries with small heaps of dung and spraying stems to deposit scent. Intruders are met very aggressively and fights are frequent. A buck is constantly looking out for does and may have several at different stages of pregnancy within his territorial area.

Reeves' muntjac are sexually mature from eight months old and do not have an annual breeding cycle as reproduction is continuous. After a gestation of seven months the doe gives birth to a single kid and is ready to mate again within a few days. At six months old kids are fully self-sufficient and leave their mother. Females are pregnant for most of their adult life and average approximately 1.5 offspring per year.

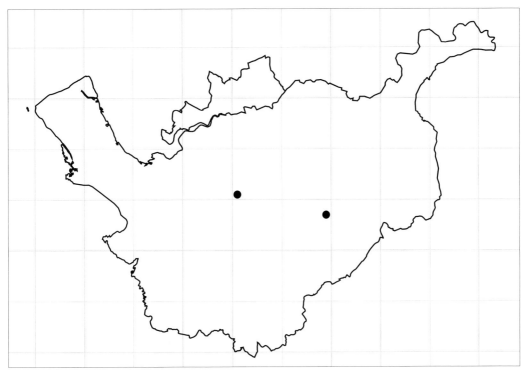

Muntjac pre-2000.

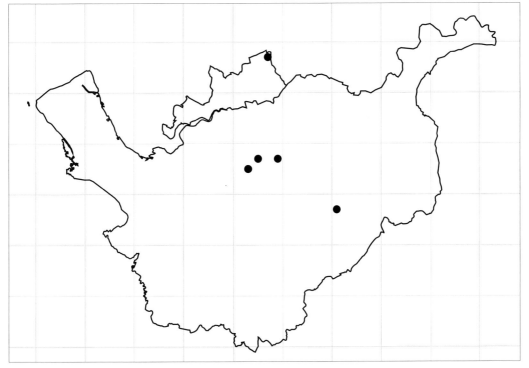

Muntjac post-2000.

Key identification features

Bucks weigh 10–18 kg and does 9–16 kg; shoulder height is similar in both sexes at 43–52 cm. Muntjac are small and appear noticeably hunched forward because their haunches are higher than their withers. When startled they run with head low and tail raised.

The coat is shiny russet brown between April and November and grey-brown over winter. The chin, throat, underside of the tail and region between the hind legs is much lighter, often white, all year round. Kids have a light-coloured spotted coat, the spots in lines with a dark brown stripe running down the back. By eight weeks old the spots disappear to be replaced by an adult coat. The broad tail is up to 15 cm long and when the deer is alarmed is held erect with the white underside displayed.

Muntjac have very large facial glands below the eyes that secrete a thick, waxy, cream-coloured substance which is used to scent-mark territories. They have a ginger forehead with pronounced black lines running to the pedicles in bucks, and a dark 'U' or kite-shaped mark in does. The hairy ears are very prominent and can appear translucent. Up to 10 cm in length, muntjac antlers are on long pedicles and curve backwards ending in a small hook. Mature males may also have a small brow tine. Antlers are shed between March and April and regrow by September.

Both sexes have long, sharp canine teeth or tusks that are capable of inflicting serious injuries when fighting. In bucks these are 2–3 cm in length and visible below the gums, but not visible in does. In mature animals the canine teeth are often broken and worn down.

Muntjac are known as the 'barking deer' due to their repeated, loud, bark-like call that can sometimes last up to an hour. An alarmed muntjac may scream; maternal does and kids squeak.

Observation

Runs and tunnels in undergrowth give a clue to muntjac presence. The territory marks of males are also characteristic: small heaps of dung and frayed stems where scent has been deposited. They also snap saplings off at about 40 cm above ground level. The most likely places to see muntjac are woodlands with a dense understorey, scrub, and overgrown gardens in urban areas. In the height of summer when vegetation is at its thickest these deer are very difficult to spot so most muntjac sightings tend to be in winter when undergrowth has died down and animals are more exposed. In poor light or dense cover fleeing muntjac can be mistaken for foxes.

Sika deer *Cervus nippon*

Order: *Artiodactyla*   Family: *Cervidae*   Genus: *Cervus*

Sika deer in Britain originate from escapees and deliberate releases from parks and zoological gardens. The species has had most success in Scotland and Ireland; current national distribution is widespread but haphazard. In England the only areas in which they seem to have flourished are south-east Dorset and the southern part of the New Forest. Their hybridisation with red deer has received much press coverage and more research is needed to get a better understanding of the origins of UK sika deer and their future here. There has been one unconfirmed sighting in Cheshire.

Sika look like a small version of a red deer but have white spots and a white rump. The tail is mostly white with a black line running down it. They are most likely to be seen in woodland, but have a taste for cultivated crops when these are available near cover. Their calls are distinctive: they emit a blood-curdling scream during the rut and a high-pitched whistle when alarmed.

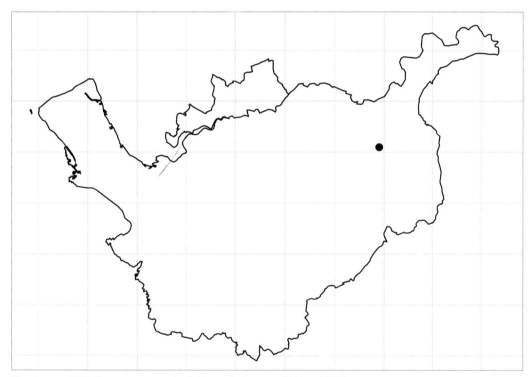

Sika deer pre-2000. There are no records of this species between 2000 and 2007.

Chinese water deer *Hydropotes mermis*

Order: *Artiodactyla*   Family: *Cervidae*   Genus: *Hydropotes*

The Chinese water deer is the least common of our wild deer and little is known about it. The present feral population derives from a number of deliberate releases and escapees, with the majority of animals still residing close to Woburn Abbey from where they are believed to have originated. These deer have a strong preference for a particular habitat consisting of tall reeds and grass close to water, especially in river flood plains and estuaries. This proclivity is likely to have restricted the species' potential to colonise more widely. They are recorded from Woburn, east into Cambridgeshire, Norfolk and Suffolk and south towards Whipsnade; small colonies have been reported in other areas, but these do not appear to persist in the long term. Isolated sightings are probably due to individual releases or escapes from private collections. There has been one unconfirmed report in Cheshire.

Chinese water deer are easily distinguished from other British deer by their large rounded ears, large eyes and the curved tusks in males. Their size is between roe and muntjac: about 50 cm at the shoulder in adults and

DAVID QUINN

weighing 8–13 kg. The hind legs are longer than the front giving a 'rump-high' appearance. The coat and short tail are red-brown in summer and sandy in winter, with no rump patch in either season. They are the only deer found in the UK that do not have antlers.

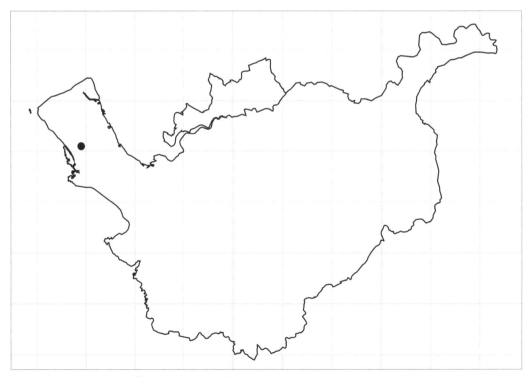

Chinese water deer pre-2000. There are no records of this species between 2000 and 2007.

11. Whales, porpoises and dolphins
Order Cetacea

Twenty-eight species of cetaceans have been recorded in waters round the British Isles (Evans, 1992). Thirteen of these species have occurred in the Mersey, Dee and East Liverpool Bay over the last 150 years with harbour porpoises, bottlenose dolphins, bottlenose whales and common dolphins being the most sighted or stranded. Less frequent visitors include Risso's dolphin, the minke whale, orca (the latest in 2001), long-finned pilot whale and fin whale. There have also been strandings of one striped (1991) and three white-beaked dolphins (in 1862, 1911 and 1989).

The number of live sightings has increased dramatically in recent years. Summer is the best time to mammal spot from land, when harbour porpoises and bottlenose dolphins tend to move inshore and into estuary areas. A visit to New Brighton, a trip on the Mersey Ferry or a stroll along the dockside may be rewarded by the sight of a fin, or even several, cutting the water's surface. Ferry journeys out of Liverpool to Ireland and the Isle of Man provide opportunities to see a greater variety of species, and an even better chance of a good view. For those less adventurous, or poor sailors, the north end of Hilbre Island is a good spot from land to look out for these elusive and appealing creatures.

Harbour or common porpoise *Phocoena phocoena*

Order: *Cetacea*   Family: *Phocoenidae*   Genus: *Phocoena*

DAVID QUINN

Status and history

This is the smallest cetacean found in British waters at just under 2 m in length and the most common to be sighted or found stranded around the Cheshire and Merseyside coast. Occurrences here have increased markedly since the 1980s, after a lean period during the middle of the twentieth century when few instances were recorded.

Going back to the late nineteenth and early twentieth century, records show the harbour porpoise was a more regular visitor, occasionally in great numbers. In 1889 a shrimp fisherman, John Hanmer, reported '... a shoal of porpoise extending a full three miles ...' passing the position of what is now the Bar Lightship in Liverpool Bay. Coward in 1910 noted the species as being '... common in Liverpool Bay and

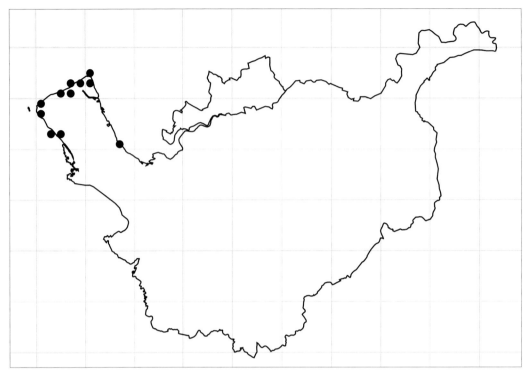

Harbour porpoise pre-2000.

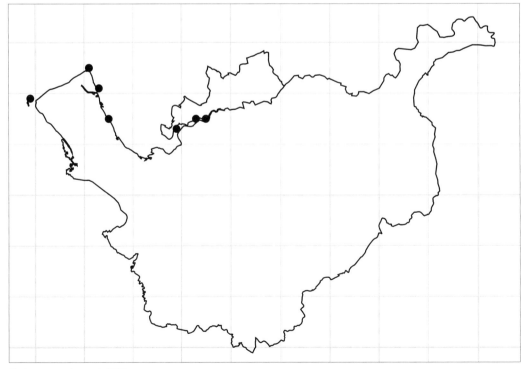

Harbour porpoise post-2000.

often ascends the estuaries' and '... occurring often in the Mersey beyond Eastham ...'.

The numbers of sightings and strandings are rising. In 2004 alone there were 10 sightings of live animals and six recorded strandings, which is roughly 10 per cent of all occurrences since the first in 1863, in a single year! A sighting of several fins off Pier Head, Liverpool in May 2006 was witnessed by many spectators and caused huge excitement. Also reported was a single, regular visitor to the Egremont Buoy off the esplanade, where it played during evenings in the summer.

UK and worldwide distribution

The species likes inshore waters, especially in the summer. It likes the cold temperate waters of the Northern Hemisphere and stays within that range, moving south in the winter and further north in the summer, probably in tune with movements of fish shoals. Squid, fish and shellfish are on the menu, but herring, sprat and sand eels constitute the main diet around the British Isles.

Description

The harbour porpoise is the most commonly seen of the species and is easily recognised as it has a low triangular dorsal fin and lacks a beak. It is small in comparison to other porpoises and has a plump body with a dark grey to bluish-coloured back, a pale belly and a rounded head. In adults, there is often a striking grey line from just below the eyes to the anterior insertion of the flipper. Flippers, tail flukes and dorsal fins are all dark grey to blackish in colour.

The pods are small, usually from two to 12 animals. A porpoise's reproductive life is extremely short as animals have a life expectancy of only 10 years and are not sexually mature until the third or fourth year. Females give birth to just one calf annually which, in addition to the short life span and reproductive window, makes local populations highly susceptible to sudden decreases in number. Calves are born around 70 cm long and have a strong bond with their mother; males play no part in the upbringing of the offspring. Weaning can occur quickly and individuals as young as six months have been known to become independent.

Key identification features

Harbour porpoises have a distinctive snub muzzle, dark back and light belly. As it is shy and retiring, the harbour porpoise can be difficult to spot and rarely shows more than a small, dark triangular fin and a glimpse of an arched back above the surface of the water. It can however be heard from some distance away making a snorting noise as it blows, giving rise to its nickname of 'puffing pig'.

Observation

Harbour porpoises do not have the flamboyant characteristics of, for example, bottlenose dolphins. When 'travelling', individuals stay low in the water so providing only a glimpse of their dark back and triangular fin as it arcs over the surface. This behaviour, combined with the poor sea states that are common in Liverpool Bay, make the species very hard to spot from land in this area.

Short-beaked common dolphin *Delphinus delphis*

Order: *Cetacea*   Family: *Delphinidae*   Genus: *Delphinus*

DAVID QUINN

Status and history

This species has been recorded occasionally in Liverpool Bay, mainly through reports of stranded animals. The records are nearly all from the latter half of the twentieth century and most have been in the last 20 years. Only two live sightings were recorded in this period; one in 1963 describes a school of 20–30 observed for several hours in Hilbre Swash, off the north coast of the island. The swash is an area of deep water, which is by far the most attractive point for dolphins along the coastline of South Liverpool Bay.

Since 2001, there have been five confirmed live sightings of the species in the northern and eastern area of the Irish Sea, with the nearest to Liverpool Bay being reported as occurring to the west of Blackpool.

UK and worldwide distribution

The short-beaked common dolphin is one of the most widely distributed species of cetaceans, as well as one of the commonest worldwide. It occurs mainly offshore in the open ocean and prefers the warm temperate waters that stretch north from the Iberian peninsula to the Faroe Islands.

Around the British Isles, this dolphin can be found in the western approaches to the English Channel and the southern Irish Sea. This is the area of the Celtic Deep, where estimates have previously put the population at around 75,000 (SCANS, 1994). Commonly found to the west of Ireland, and off the edge of the continental shelf in general, it also occurs around the Inner Hebrides up to Skye and, in recent years, further north and east around Shetland and Orkney, and in the northern North Sea. This may be related to global warming and the changes in the Gulf Stream, making these waters more attractive for the species. It is rarer on the east side of Britain.

Description

There are two species of common dolphin, the short-beaked and the long-beaked (*Delphinus capensis*), but only the short-beaked variant is found in waters around the UK.

A very sociable animal, the common dolphin is often found in large schools. However, schools spotted off the British coast are comparatively small, with less than 30 or so animals. They may also be seen in pairs, and even occasionally alone. Groups will hunt co-operatively by herding fish together then feeding off the frenzied mass of the shoal. Their diet is very varied including cod, hake, mackerel, sardine, pilchard, horse mackerel, sprat, sand eel and herring, as well as squid.

They communicate with squeals that can be clearly heard and tend to vocalise often. This presumably aids in the formation of the strong social bonds that are evident, particularly between mother and baby. Common dolphins mainly give birth in the spring and autumn after a gestation of around 11 months. Other females are known to help both with the birth and with babysitting duties so that mothers may feed in peace.

Common dolphins can live 30–35 years in the wild. They are mature at around six years old and, once sexually mature, a female will give birth about every two to three years.

Key identification features

With a length of 2.1–2.4 m and weight of 75–85 kg, a fully grown short-beaked common dolphin is one

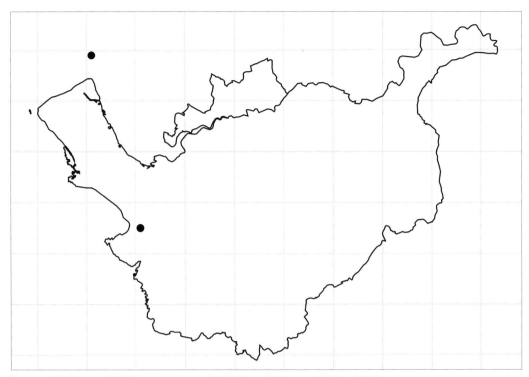

Common dolphin pre-2000. There are no records of this species between 2000 and 2007.

of the smallest of the true dolphins. The body and the beak are long and slender, and the dorsal fin is tall and pointed. Nicknames include the 'hourglass' and 'criss-cross' dolphin, owing to the distinctive hourglass pattern on its flanks with yellow or tan at the front and a light grey behind.

The other distinctive feature is behaviour. Not only are dolphins agile and boisterous, often engaging in aerial acrobatics, but they also have a reputation for bow-riding, often accompanying a boat for miles and swimming with great speed and stamina.

Observation

This is an offshore species and sightings from aerial surveys in the area of Liverpool Bay have been few (SCANS, 2005), so it is unlikely that these dolphins will be spotted from the shores round the Bay. Far better opportunities to see these attractive animals may be had on the ferries that cross the Irish Sea to Ireland and the Isle of Man, as well as those heading out towards the Spanish and French coasts.

Northern bottlenose whale *Hyperoodon ampullatus*

Order: *Cetacea*   Family: *Ziphiidae*   Genus: *Hyperoodon*

DAVID QUINN

Status and history

Northern bottlenose whales are described as being rare in the Irish Sea, and are not included in the review of cetaceans in Liverpool Bay (Evans & Shepherd, 2001). Despite this, they are the third most numerous species noted over the last 170 years in this area. Unfortunately, most of these instances have been stranded animals.

The records for the Cheshire, Wirral and Merseyside coasts show an interesting pattern. Most records date back to the nineteenth century, with the first noted in 1829. This particular whale was found stranded at the mouth of the Mersey and recorded by London's *Magazine of Natural History*; its skeleton is now in Liverpool Museum. There were 14 records spanning from 1829 to 1881 with the majority of these (eight) being live animals, at least when first sighted. Their fates were recorded as 'killed', 'captured', 'shot' and, more luckily, 'escaped'! Records fall off in the twentieth century, with a further six spread out across the years, the last being a stranding in 1998. They have never been regular visitors to these shores, which, given this kind of reception, seems hardly surprising.

UK and worldwide distribution

The species is found in the northern Atlantic. These whales like cold waters and tend to gravitate farther north in temperate and subarctic waters. They are not known to migrate to warmer waters. Bottlenose whales are offshore animals, found mainly over deep water canyons of over 1,000 m. In waters around the UK and Ireland, the species is sighted primarily in the Faroe-Shetland Channel and Rockall Trough, and has occasionally been seen in coastal waters, for example around the Isle of Skye or off northern Scotland. A

few individuals do wander off the beaten track, for example in January 2006, when one found its way up the River Thames into the heart of London. This is perhaps the explanation for our occasional visitors.

Description

Bottlenose whales live in small social groups of two to four individuals, but rarely more than 10. The groups appear to comprise different sexes and ages, with mature males tending to separate from female groups after the mating season. Gestation is about 12 months, with most calves being born in the spring and summer. It is thought that bottlenose whales can live for 40 or even 50 years.

They routinely dive to 800 m and can go to depths of at least 1,500 m, but do not seem to travel far when diving deep. Although they sometimes remain underwater for up to two hours, dives of 15 minutes to one hour are more common. Once back up they will remain on the surface for over 10 minutes, puffing and blowing their bushy spouts. They sometimes spy-hop, tail-slap and breach, in a similar way to dolphins.

This species is known to be inquisitive and whales will approach boats readily. They are also known for refusing to abandon injured or stranded members of their pod. Such traits made them an easy target for whalers and tens of thousands were slaughtered at the height of commercial whaling. Those same characteristics now make them an appealing species for whale-watching expeditions.

Bottlenose whales' penchant for deeper waters, where they find their main prey species *Gonatus fabricii*, a deep-water squid, makes the shallow Mersey coast an unattractive prospect for them.

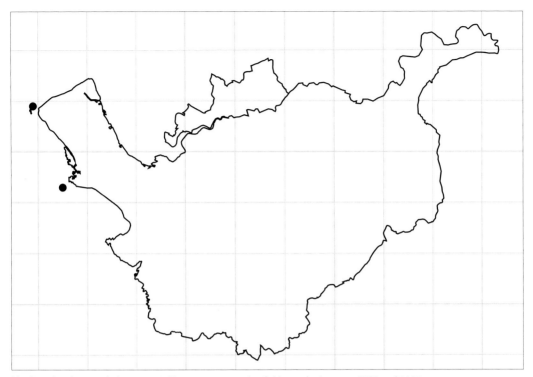

Northern bottlenose whale pre-2000. There are no records of this species between 2000 and 2007.

Key identification features

Nicknames include 'bottlehead' and 'steephead' due to an enlarged, bulbous head, which contains spermaceti oil. Beneath this bulge is a short, pointed beak with two small teeth in the lower lip in the males. Their colour varies from a chocolate brown through creamy brown to yellow, depending on the growth of diatoms on the body. The bulbous head often has a very light patch, particularly in the males.

Fully grown, an individual can reach 9 m in length and weigh 7.5 tonnes. The dorsal fin is quite prominent, about two-thirds of the way down the back with a slightly falcate or concave trailing edge. A distinctive feature is the spout, which is high and bushy, and directed slightly forward.

Observation

You will be very lucky to spot one of these creatures from the shore, unless it is the unfortunate instance of a stranded whale. A better option would be the view from a boat, although the whales' attraction to the colder, northern areas and the incipient effects of global warming would make a sighting more likely on a cruise to the Orkneys or Shetlands than to the Isle of Man.

Bottlenose dolphin *Tursiops truncatus*

Order: *Cetacea*   Family: *Delphinidae*   Genus: *Tursiops*

DAVID QUINN

Status and history

The bottlenose dolphin is an irregular visitor to the shores round the Mersey and the Dee, with just over 40 occurrences noted since records began 170 years ago. The first record appears in 1918, sighted by Eric Hardy, a reliable source, but his next encounter was not until 1942. Regular recording began at Hilbre Island in the 1960s and, since then, numbers of sightings have shown a steady climb from four in the 1960s, to eight in the 1980s and 11 in the 1990s.

Most sightings of live animals are from Hilbre Island, where earlier records show pods of 10–20 individuals, as opposed to the much smaller numbers of two to four per pod in recent years. Although this appears to be the best spot locally for sightings, animals have occasionally popped up at West Kirby, Parkgate, Seaforth and Formby Point, with one sighting off the Mersey Ferry in May 2000.

Strandings are rare, and there have been none recorded along the Cheshire, Merseyside and Wirral coast since 1968 at Formby. This inshore species copes well with the shallow waters and shifting sands of the area, where risk of stranding is high.

UK and worldwide distribution

Bottlenose dolphins are found all round the coast of Britain and there are resident populations in the Moray Firth in Scotland, Cardigan Bay in Wales and on the south-west tip of Britain. The latter population returned in 1991 after a recorded absence of 20 years.

Although they are ubiquitous around the coasts of the world, bottlenose dolphins become rarer farther to the south and north, and are not present in the polar regions. Whilst they are known to prefer inshore areas in depths of as little as one to three metres, some are also found far out from the shore. This varying behaviour has led to some speculation as to the presence of inshore and offshore ecotypes. The Cardigan Bay dolphins, which may well be the source of our visitors, seem to cope well with the shallow waters in the Bay, but are known to venture out to the deeper areas of the St George's Channel and Celtic Deep to forage.

Description

The resident population in the Moray Firth shows a regular yearly pattern of movement, moving inshore and up the estuary in spring. Similarly at Hilbre, most sightings have been between March and October; this may be linked to both feeding and birthing behaviours. Females give birth to single young, and feed them with rich milk that has four times the fat of human milk. The calves can swim immediately and grow quickly. It is believed adult females can live to 50 years of age in the wild; they do not do anywhere near as well in captivity.

Understanding their feeding habits and the local tidal conditions can provide the best chance to observe dolphins. They search out tidal streams where prey such as fish and squid collect and, swimming against the current, they move into the moving mass of prey to pick off their food. They are also known to hunt in groups, working as a team to corral fish into a massive ball and then feed on the trapped shoal.

Bottlenose dolphins may be found in pods (or schools) of 300–500 individuals. They are rarely solitary and it is the single animals that seem to

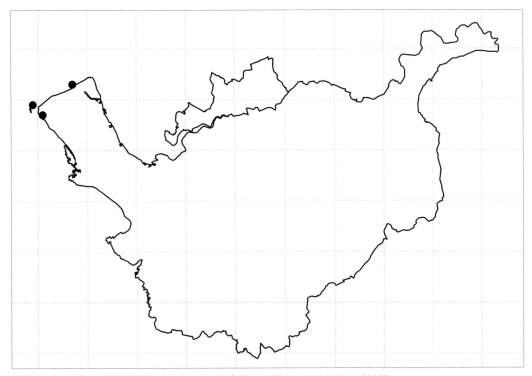

Bottlenose dolphin pre-2000. There are no records of this species between 2000 and 2007.

seek out human company. Like many social animals at the higher end of the evolutionary scale, they are considered to be intelligent. They communicate, as do many cetaceans, by a form of sonar which is evident when they are hunting, commuting and during social interactions. Mothers, for example, whistle to their newborn and the calf recognises and responds to its mother's call even in a large mixed pod.

Their attraction to humans is well documented, and the recent trend of swimming with dolphins is associated with these animals. In actual fact they can be aggressive and potentially dangerous if approached, particularly if young are present. They are now known to be the culprits in a spate of deaths of harbour porpoise, actively attacking them, tossing them into the air and causing appalling internal injuries that are often fatal. The smaller, more even-tempered common dolphin is probably a more appropriate swimming buddy.

Key identification features

The size of bottlenose dolphins varies hugely, with adult weights from 150 to 650 kg and lengths ranging from 1.9 to 4 m. The range is epitomised by comparison of the slender bottlenose dolphin resident off California to its robust Scottish cousin on the Moray Firth, where the harsher climate requires a little more padding. Size variation is also believed to be associated to the aforementioned ecotypes, with the larger animals being found further from shore.

Colour does vary but generally they are grey-blue on the back with a lighter belly. The dorsal fin is almost central on the back and has a distinct falcate trailing edge. The face has a characteristic blunt beak, with a slightly protruding lower lip and an upward curl at the mouth edges, giving the appearance of a smile. Behaviour is often flamboyant: breaching, tail slapping and leaping out of the water, sometimes to several metres high.

Observation

Hilbre Island is the best place for spotting these animals, although the sea there is often choppy making observation difficult unless they breach out of the water. It can be other animals that give away their presence: a collection of sea birds and other marine wildlife may indicate the position of a pod as they compete for food. Changes in tidal streams, indicated by a demarcation line on the surface, and a change in the colour of the sea is another area to look out for these attractive and entertaining animals.

12. Escapes and exotics

Since humans first began to move around the world around 8,000 years ago they have been taking animals with them and introducing these animals into habitats and ecosystems where they did not previously occur. Mammals have been introduced by accident, for food, pest control and visual appeal and have been released or escaped from commercial and domestic situations. Initially they may have been regarded as aliens but over time many have thrived and become accepted.

Of the free-roaming terrestrial mammal species found in Britain, just over half are 'native'. Several of the introduced species such as Reeves' muntjac are increasing in range and abundance. Some introduced species such as fallow deer and rabbits have been here so long that we treat them as naturalised. In some instances the genealogy of British species is mixed. This is the case with roe deer in the south of England, whose ancestors came from the continent in the nineteenth century whereas Scottish roe deer probably descend from a population remaining after the last Ice Age. More recent arrivals include the American grey squirrel and the American mink.

There are many cases worldwide of damage to the physical environment and to native fauna and flora as a result of non-native species introductions. In Great Britain examples include the grey squirrel, the New Zealand flatworm (*Artioposthia triangulata*) and Japanese knotweed (*Fallopia japonica*). It is important to control and regulate releases of non-native species to ensure that further damage to the environment does not occur.

It is interesting to note that if an individual of an alien species is captured, it is an offence to re-release it without a licence. For example, in the UK it is illegal to release a captured grey squirrel, even if the animal was found injured and nursed back to health or a youngster reared in captivity.

The general legislation for the control of releases of non-native species in Britain is provided by the Wildlife and Countryside Act 1981, Sections 14–15 and Schedule 9. In England and Wales, further amendments to the enforcement procedures are covered by the Countryside and Rights of Way Act, 2000, Schedule 12.

The regulations of imports of certain mammals are covered by the Destructive Imported Animals Act 1932. The Dangerous Wild Animals Act 1976 and the Zoo Licensing Act 1981 contain provisions to prevent the escape of captive non-native species.

Some of the mammals we today regard as aliens would not have been heard of a century ago. There is a good chance the future may bring us more mammals on a permanent basis as climate change enables and encourages some species to establish themselves.

Exotic mammals in Cheshire

Siberian chipmunk (Tamias sibiricus)

Chipmunks of the genus *Tamias* are found in the deciduous forest areas within eastern Canada and North America. Today they are increasingly being kept as pets. This small burrowing rodent is fast and active, jumps, climbs and has a tendency to try to escape from cages. Larger than the European or Asian chipmunk but noticeably smaller than the red squirrel, they have a body length of 130–190 mm and a tail length of 75–130 mm. They are omnivorous, eating grain, nuts, birds' eggs, fungi, worms and insects.

Reports of chipmunks have come from Moore Nature Reserve in Warrington. The ranger there reports that chipmunks (possibly Siberian) were present having been intentionally (and illegally) released on 14 June 2006. They were spotted around for a few weeks before presumably dying out. Recorded sightings were made on 16, 23 and 30 June 2006.

Chipmunks have also been reported and photographed at Haddocks Wood in Runcorn in 2007. Again, these animals were almost certainly illegally released.

Bennett's or red-necked wallaby (*Macropus rufogriseus fruticus/rufogriseus*)

DAVID QUINN

These animals are probably the only wild marsupial mammals in the UK. They can cope well in Britain as there is a similar climate in their native Tasmania. They are adaptable feeders, grazing on grasses and browsing on available plant material and in Britain have come to favour heathland and grassland habitats. Bennett's wallabies are not large animals, growing to a maximum of around 80–100 cm tall. They are grey-brown in colour with red-brown neck, and a pronounced reddish tone to the fur on their shoulders and rump.

Originally imported for zoos and private collections around the turn of the twentieth century, Bennett's wallaby escapees have successfully established breeding populations and have survived in naturalised groups since the 1940s. Wild populations

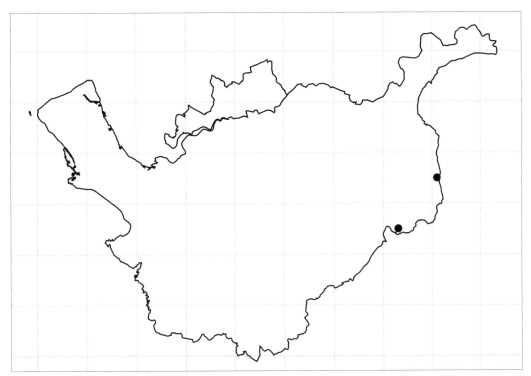

Bennett's wallaby pre-2000. There are no records of this species between 2000 and 2007.

have established in Scotland, the Derbyshire Peak District and in mid-Sussex, the latter dying out in the 1970s. The Peak District animals came from a private zoo at Roaches House near Leek: five wallabies escaped when fencing was not properly maintained during the Second World War. In 2008 there was just one 12-year-old female left, with no males recorded since 1996. No problems with the remaining populations have been recorded as yet, but little research has been carried out.

Big cats

Since 1976 there have been numerous sightings of large felines in the UK. A number of agencies take such instances seriously, especially as many sightings have been by professionals such as the police. Evidence has been from spoor, fur samples and footprints along with amateur videos and photographs.

In 1976 the Dangerous Wild Animals Act was passed requiring licensing of large carnivores and it is likely that many animals kept as exotic pets were released from private collections as a result of this legislation. To be still present in the wild today, released animals must have bred, although life span can be 30–40 years. The most likely candidates are puma or leopard (black leopards are known as panthers).

Moorland and areas with low population density are favoured and farmers have reported finding mutilated and partially eaten sheep in areas where sightings have occurred. Sightings have also been reported from suburban areas.

The big cat keepers at Chester Zoo consider that there are many exotic cat species that *could* survive in the UK, and that their behaviour is such that they *might* go undetected for many years while feeding on small prey such as rabbits and invertebrates. However, it is unlikely that there are many, if any, cats out there in reality. It is probable that a released cat would try to keep away from people as far as possible, moving to remote areas and keeping very well hidden. Most reports turn out to be large dogs or foxes, whose movement can be very cat-like.

Escapes from zoos

It is clear that historically zoos have been responsible for some of the introductions to the UK of species such as the muntjac, wallaby and Chinese water deer. It is reassuring to know that legislation and management practices have now changed dramatically and escapes from zoos are extremely rare, being treated far more seriously than in the past.

Escapes from Chester Zoo

Old records show that the North American beaver (*Castor canadensis*) did escape from Chester Zoo in the 1960s and again in 1985. In the earlier case it is not clear if the animals were recaptured, but there were no sightings recorded. Of the pair that escaped in May 1985, one was recaptured at Ellesmere Port Boat Museum the following day, but died in the zoo soon after due to suspected poisoning from oily water. The other was spotted in the Shropshire Union canal the day after the escape but was not seen again and presumed dead after that point.

In June 1963 two raccoons (*Procyon lotor*) were received at Chester Zoo from Dallas Zoo. One of them escaped the following day and was spotted in the local area a couple of times. Each time the individual evaded capture by disappearing down drains. The other animal survived at the zoo until 1969.

Officers at DEFRA who deal with reports of alien animals have not had any reports for Cheshire or Shropshire for many years. They used to visit Chester Zoo to inspect the coypu (*Myocastor coypus*) exhibit to ensure the animals could not get out; they never did, and are not kept at the zoo any longer.

Possible escapees

The range of exotic mammals kept as pets today means that the chances of exotic escapees is quite high. Chinchillas, degus and gerbils, raccoons, skunks and even sugar gliders may be resident in Cheshire homes. The chances of populations actually establishing in the county though is much smaller.

Exotic farm stock is also popular today and animals such as wild boar, llama or alpacas might escape from farms or from vehicles. The message is to keep your eyes open and don't immediately discount unusual sightings.

On a UK scale, reintroductions of animals such as beavers are taking place. In these circumstances, the ecological case for reintroduction is strong and such projects are being carefully controlled and assessed.

13. Conservation action in Cheshire

Cheshire region Biodiversity Action Plans

Biodiversity action planning in context: international and national biodiversity conservation

The United Nations Conference on Environment and Development (UNCED), also known as the Rio Earth Summit, was held in 1992 and signified the first global acceptance of our responsibility to safeguard the environment. The Convention on Biological Diversity formed an important part of the summit resulting in an agreement between 153 countries worldwide, including the United Kingdom, to 'develop national strategies, plans or programmes for the conservation and sustainable use of biological diversity.'

In response to this agreement the UK Government launched *Biodiversity: the UK Action Plan* in 1994 and formed the UK Biodiversity Steering Group. Habitats and species had suffered severe decline during the twentieth century due mainly to the combined effects of agricultural intensification, land loss to development, pollution and inappropriate management. The Steering Group was tasked with identifying habitats and species of conservation concern both nationally and internationally. Those habitats and species most at risk were highlighted as priorities for which Action Plans were written to set out the reasons for decline and threats, alongside recovery targets and prioritised work programmes. The UK Biodiversity Action Plan (UK BAP) was reviewed in 2006–07 during which time the priority list was updated and quantitative targets set. The results of this review, as available at the time of writing, are included in Appendix 1. The full list of UK species, habitats and local action plans can be found at www.ukbap.org.uk.

Regional and local biodiversity conservation

In 1996 the newly formed Regional Biodiversity Steering Group commissioned a biodiversity audit of north-west England. The audit brought together all available information on the status of habitats and species in the region allowing priorities for conservation to be identified. It also informed the production of Local Biodiversity Action Plans (LBAPs) including the Cheshire region LBAP completed in 1997. The Cheshire region Biodiversity Partnership (CrBP), consisting of over 40 relevant multisector member organisations, was established to drive forward local biodiversity conservation within the administrative areas of Cheshire County Council, Wirral Council, Halton Borough Council and Warrington Borough Council.

The Cheshire region LBAP can be viewed at www. cheshire-biodiversity.org.uk. Over 70 action plans have so far been written for habitats and species in the Cheshire region that are either identified as priorities in the UK BAP or are of particular local significance. Each action plan identifies the current status of the habitat or species, the reasons for its decline and the prioritised actions needed to achieve targeted recovery.

Local biodiversity conservation action towards these targets is achieved through a number of mechanisms. Documents such as *Halton's Biodiversity Action Plan* and Warrington Borough Council's *Nature Matters* provide a framework for protecting and enhancing biodiversity within each local authority area. Similarly the Wirral Biodiversity Group exists to oversee delivery of actions specific to the Wirral peninsula including taking the lead on all coastal and marine action plans. Other local authorities including Vale Royal and Macclesfield have produced biodiversity audits to help recognise and protect important areas of their natural environment. In addition, various action groups are well established to support a number of action plans, bringing together experts and delivery partners to advise on best practice and monitor progress towards targets.

Reporting and monitoring local progress towards national action plan targets is particularly important.

Conservation practitioners are increasingly using the 2004 Biodiversity Action Reporting System (BARS) to publish actions proposed, under way and completed towards achieving biodiversity targets. The BARS system also allows status, trends and losses to be recorded and provides a searchable database of biodiversity information to the general public.

Local delivery partners

Mammal conservation in Cheshire is driven forward by a number of key groups and organisations. Members of the CrBP have joined forces with the Cheshire Mammal Group to form species-specific mammal action groups to ensure delivery and accurate reporting of achievements against each plan. CrBP members such as Cheshire Wildlife Trust, Chester Zoo, Natural England, the Environment Agency, British Association for Shooting and Conservation and a number of local authorities are particularly active. Other specialist organisations such as the Cheshire Bat Group, Liverpool Museum, the Mersey and Dee Estuary Conservation Groups and the Wirral Biodiversity Group provide further expertise and voluntary support.

The North West Dormouse Partnership works cross-border between Cheshire and North Wales bringing together expertise from the Cheshire Wildlife Trust, Chester Zoo, North East Wales Wildlife Trust, Forestry Commission, Countryside Council for Wales and Environment Agency Wales to further the conservation of dormice through regular monitoring and research.

rECOrd provides an invaluable data collection facility that helps inform conservation action for mammals in the region.

Cheshire region mammal action plans

The CrBP has published nine mammal action plans for the Cheshire region for the following species: bats, brown hare, dormouse, otter, water vole, small cetaceans, grey seal, harvest mouse and polecat. These plans contain targets and actions for 14 UK priority species and a further eight species of local importance. It is worth noting that the geographical area covered by the CrBP does not tally exactly with the area covered by Cheshire Mammal Group and the maps included in this atlas. The nine mammal action plans can be found in Appendix 1.

The return of hazel dormice to Cheshire

Background

In the early 1990s it became obvious that the dormouse was in decline. The Mammal Society carried out a national survey based on nut hunting in the 1980s; this survey was repeated in 1993/94 and became the largest voluntary wildlife survey undertaken in Europe. The results showed dormice had disappeared from much of northern England and that their numbers were still in decline.

Historically, the biggest threat to dormice has been loss and degradation of their woodland habitat. Dormice need large areas to survive: it is estimated that at least 20 hectares of suitable habitat with diverse tree and shrub species is required to support a population long term but there are few broad-leaved woodlands of this size left in Cheshire. Fragmentation of the remaining woodlands is also a problem. As dormice are reluctant to come down to ground level, and will not cross open spaces, they are poor colonisers and will not reach isolated woodlands. They will travel along hedges, but the lack of management

leading to gaps in or complete loss of hedgerows, has further reduced their ability to move around.

English Nature (now Natural England) initiated the Species Recovery Programme in 1991, with the aim of achieving the long-term self-sustained survival in the wild of species of plants and animals under threat of extinction. The dormouse was added to the programme and action to preserve the species had a dual focus: first, improving habitat and linking up areas of habitat in southern England where dormice were still relatively abundant; and secondly, returning dormice to a number of northern counties where they had been lost.

Start of the project

The first UK dormouse reintroductions took place in Cambridgeshire in 1993 and Nottinghamshire in 1994. Cheshire Wildlife Trust was approached in 1995 by the coordinator of the release programme, Dr Pat Morris from Royal Holloway College London, regarding potential reintroduction sites in Cheshire.

The Trust responded enthusiastically to this opportunity to bring the dormouse back to the county and appointed a volunteer dormouse project officer to run the project. Information on ecology and habitat requirements of dormice was collected and a keen group of Cheshire dormouse volunteers assembled. These people turned out to be vital to the success of the project.

Good dormouse habitat is scarce in Cheshire with less than two per cent cover of natural woodland, so finding a suitable 20-hectare site was a tough challenge. On the advice of the Forestry Commission, Stockton Dingle in the Wych Valley on the southern edge of Cheshire was investigated. The area is a steep-sided clough woodland, dominated by hazel and with few big standard trees: ideal dormouse habitat. The only drawback was its relatively small size: nine hectares at most. However, the Dingle is part of a mosaic of woodlands along the valley, the majority of which have similar vegetation and structure. These woodlands are well connected by hedgerows and trees alongside the Wych Brook. All this added up to plenty of space for dormice to colonise and thus a potential reintroduction site had been found.

Before the reintroduction

Before the release it was vital to ensure dormice were not already present. This was unlikely as the last recorded dormouse sighting in Cheshire was in 1910. However, as dormice can be very secretive they could have been overlooked. In the autumn of 1995 the site was intensively searched for dormouse-chewed hazelnuts: as there is abundant fruiting hazel in Stockton Dingle nut hunting was an appropriate survey technique to check for the presence of the species.

The dormice for reintroduction were all bred in captivity by a handful of dedicated private individuals. Ideally all the animals should have been released at the same time, but at that point the captive breeding programme was not large enough to supply enough dormice in a single year. As a result Cheshire received its dormice in two batches, 29 mice in 1996 and a further 24 in 1997.

The first dormice arrived in June 1996 consisting of 10 males and 19 females. The skewed sex ratio was to promote reproductive output in the first year since dormice are not monogamous and each male can mate with more than one female. Most animals were already in groups of three: two females with each male. Having them in social groups was important as dormice are choosy about their companions—putting two animals together does not guarantee they will be friendly!

Hard or soft release?

There are two main techniques for releasing animals into the wild. Hard release is the simplest: the animals are released directly into the environment and left to survive as best they can. Soft release provides more support: the animals are given time to acclimatise, and are provided with food, water and shelter while they explore and adapt to their new home.

Radio-tracking studies elsewhere have shown that dormice do not forage at random, instead moving directly to chosen food sources. It appears they have a mental image of their home range and know where to find food. Hard-released animals do not have this local knowledge, so are unable to forage efficiently straightaway. Soft release provides the dormice with a reliable food source while they become familiar with their surroundings. For this reintroduction, soft-release dormice were provided with large mesh cages about 45 cm square and 1 m tall. The cages were tied into hazel stools and branches were put inside so the dormice were in intimate contact with their new habitat. A nest-box was placed in each cage and a plastic tube enabled food to be provided without opening the cage. The dormice were fed daily with items with which they were familiar from captivity such as sunflower seeds, grapes, bananas and rich tea biscuits! After a fortnight a small hole was made in a top corner of each cage, adjacent to a hazel stem, so the dormice could come and go as they pleased. Daily feeding continued for several weeks, examination of the food remains each day giving an indication of whether food was being found elsewhere. Food provision was reduced gradually and discontinued in early September.

The 1996 release

Cheshire's 'new' dormice went into 10 soft-release cages spaced about 100 m apart along Stockton Dingle. Since dormouse home ranges average 400 m in diameter, the released animals had a good chance of meeting each other. The release took place in June to allow the animals time to breed in that first year. Supplementary feeding in the release cages compensated for any shortage of natural food.

This strategy proved to be the right choice with some females even starting families in the nest-boxes inside the release cages, despite the fact that they were free to come and go. The wisdom of this was further proven the following spring when very few of the captive-bred dormice were found; they were assumed not to have survived hibernation. However, animals born in that first year did survive the winter, and were thriving by the next spring.

Handling and licensing

Dormice are protected under the Wildlife and Countryside Act (1981) so anyone having close contact with them needs a licence. In the first couple of years after the Cheshire release, 'experts' helped with monitoring and trained the Cheshire volunteers so that they could apply for their own licences. Since 1998 the project has been independent, with enough licensed personnel to carry out all the monitoring, so much so that since 2000 the Cheshire dormouse team has provided training for dormouse workers in Staffordshire and North Wales.

Monitoring

As Cheshire was one of the first UK dormouse reintroduction schemes, and success is never guaranteed, the animals were very carefully monitored for several years after the release. To enable individual animals to be recognised they were identified by a unique tattoo in their ears. The tattoos were tiny and reading them was difficult, but they meant that more could be learned about each animal's movements and behaviour. The release animals were tattooed before they arrived in Cheshire. However, since it was necessary to mark new young dormice born at Stockton Dingle, volunteers from Cheshire also learned the tattooing technique.

Before the dormice arrived 120 special nest-boxes were put up in Stockton Dingle, and many of these were installed with the help of members of the British Trust for Conservation Volunteers. The nest-boxes are similar to those used for blue tits in domestic gardens. Nest-boxes are a good, if expensive, way to survey for dormice as their nests are easily distinguished from those of birds and other small mammals.

To minimise disturbance in the initial stages, monitoring in 1996 was only carried out in August, October and November. August results were encouraging: eight dormice were found, including an unmarked juvenile. This youngster was probably from a litter born in a release cage soon after the reintroduction (its mother may have been pregnant before arriving in Cheshire). Of the others, two females were pregnant, and another two had litters of young babies.

By October a large number of empty nests in nest-boxes indicated the dormice had been and gone! Six of the eight animals found were juveniles providing further evidence of a successful breeding season. In November 11 animals were found and most of them were unmarked (so born after the release). They were obviously preparing for winter hibernation, as there were some very heavy animals: one tipped the scales at 29 g.

The 1997 release

The second release phase in 1997 was dependent upon some of the 1996 animals surviving their first winter. Monitoring in April 1997 turned up a single dormouse but this was enough for the 1997 release to proceed. In spring 1997, in preparation for the release of more dormice, additional nest-boxes were put up towards the northern end of Stockton Dingle and into an arm of woodland to the west, bringing the total number of boxes to 280. The release cages from 1996 were still in place, but some were moved to avoid introducing new dormice into the territories of established males found during surveys later that spring.

The 1997 release comprised 24 animals: 11 males and 13 females. These were put in cages as 11 male-female pairs, and two females together placed near where a lone male had been found in June. The same pattern was followed as the year before, with daily feeding for three weeks after the release. A gradual reduction in feeding followed and was finally discontinued at the end of August.

The week the dormice arrived, towards the end of June, was cold and wet. In some cages no food was taken for several days and the dormice were assumed to be in summer torpor due to the cold. Also, aggression problems were observed in some cages and it became apparent that not all pairs were compatible. The discovery of several animals looking 'the worse for wear' prompted the addition of extra nest-boxes in all the cages where signs of aggression were seen.

In 1997 the dormice appeared to adapt swiftly to their new surroundings. During the first week several animals were active during the day, but thereafter they became more nocturnal and were rarely seen. One female was seen in a cage several weeks after the release and was clearly lactating.

After the release the dormice in some cages disappeared very quickly. These animals may have dispersed away from the cages, been chased away by resident wild dormice, been predated or died of other causes. By the time feeding was discontinued in 1997 food was only being taken from four cages, and only in small quantities from three of these. In the autumn 1997 monitoring period, just one of that year's released animals, a male, was found. However, there were 29 captures of adults and juveniles, comprising 20 different individuals, and five of these were animals released in 1996. In addition six litters of babies were seen, with one female having two litters. Eleven new juveniles were tattooed, and three more juveniles were found which were too small to tattoo. It appears they had had another good breeding year and the reintroduction was looking positive.

Staffordshire dormouse project

In the spring of 1999 Cheshire Wildlife Trust received a phone call from a farmer on the Cheshire-Staffordshire border, who was sure he had seen a dormouse in his wood. The description he gave sounded good: a small golden-brown animal with big black eyes, which had stared him in the face before escaping up a tree. This was very exciting as the wood was in Staffordshire, where dormice were supposed to be extinct.

A site visit revealed a small wood that appeared initially unpromising as a dormouse habitat—part ancient woodland, part larch plantation and part birch-dominated regrowth. However, the farmer's description was convincing, so 20 nest-boxes were put up in his wood. A few months later two torpid dormice were found, the first recorded in Staffordshire in living memory. Old maps showed that this wood was a remnant of a much larger forest covering an area around Loggerheads. Was it possible that dormice survived in other woodland fragments? Around this time the Staffordshire Wildlife Trust and the newly established Staffordshire Mammal Group became involved and they now monitor these sites.

Progress to date

Over the years the Cheshire dormice have done well. Tattooing was discontinued after May 2000 as much information had already been gained including the discovery that a couple of the Cheshire individuals had reached four years of age. There was no spring monitoring in 2001 due to the outbreak of foot-and-mouth disease and the July and August monitoring was discontinued.

Dormice seem able to adjust their breeding in response to environmental conditions, especially food availability. In Cheshire, observations in 2004 and 2006 noted abundant fruiting of hazel, hawthorn, blackthorn and crab apple, all dormouse food sources. The dormice responded by producing much larger litters of young. In 2004 litters of between six and nine youngsters were recorded.

In recent years dormouse nest tubes have become available. These are made of corrugated plastic with a wooden insert and are much cheaper than wooden boxes, as well as being lighter to carry and easier to install. In Cheshire, nest tubes are now the principle means of surveying new woodlands for dormice.

Volunteers have been vital to the success of the project, and over the years dozens of people have been involved. Something about dormice seems to attract long-term commitment and, as a result, some volunteers have become highly skilled and knowledgeable.

The future

Since the 1996–97 dormouse reintroduction in Cheshire, regular monitoring has shown a gradual increase in both dormouse population and the area occupied. It has helped increase understanding of dormouse ecology and it is hoped that monitoring will continue to further record progress in the long term.

Since the Cheshire reintroduction, our knowledge of dormouse habitat requirements has changed. At the start of the project, it was thought they could only survive in ancient woodland; it is now known that they can thrive in secondary woodland, mixed woodland and even conifer plantations. Thus there are many more locations in Cheshire which might support populations of dormice. Cheshire Wildlife Trust hopes to survey as many of these as possible, in the hope of uncovering a previously undetected native population or another site that could be considered for a reintroduction. The reintroduction programme is now managed by the People's Trust for Endangered Species.

It is likely that climate change will have a big influence on dormice in the future, although the effects are uncertain. If warmer summers result in more abundant fruiting of dormouse food plants this will be beneficial. However, warmer winters may adversely affect hibernation. Dormice cannot drop their body temperature below the surrounding air temperature, so in warmer winters their metabolism will not slow so much and they will use their fat reserves faster. Warmer springs may result in dormice emerging from hibernation earlier but, if there is no food available that early in the year, they will starve. As with other species, ongoing monitoring is vital to help understand the effects of climate change.

Chester Zoo Harvest Mouse Project

Introduction and background

It was almost by chance that Chester Zoo acquired its first harvest mice in 1986, when Glasgow University and London Zoo were looking to rehome their remaining stock of this species. Regarded at that time as challenging to keep in captivity owing to their need for intensive management, many collections were finding it difficult to maintain populations of harvest mice over the long term. From the small founder stock of five animals, supplemented regularly with new blood from across the UK, Chester Zoo developed a highly successful captive breeding programme. Since 2002 this programme has provided more than a thousand mice for release projects and educational displays.

The research project

The fundamental aims of the Chester Zoo Harvest Mouse Project are:

- Husbandry guidelines: to assist others keeping this species in captivity, and learn about and improve welfare conditions
- Reintroduction protocol: to establish a protocol for the reintroduction of this species, applicable across its range, should that ever be a conservation requirement for harvest mice in the future

The first aim was achieved through a captive breeding programme, and the second through a series of release trials on land at Chester Zoo and selected sites in Cheshire.

Captive breeding programme

The captive breeding programme has been a real community affair involving many volunteers from the local area. Zoo members, Cheshire Wildlife Trust members and keen local naturalists have all been included, successfully breeding mice in their own homes. Initially, these 'satellite' breeders were given instructions and training but, as their experience grew, they provided feedback into the project and increased understanding of the behaviour and needs of the animals. The husbandry guidelines, published in 2000, developed from this work.

Breeding harvest mice can be a tricky business as they need close observation, keen record keeping and are very quick when being handled! As mice bred, young animals were returned to the zoo for re-pairing into new groups or sending on to other collections.

A detailed stud book, recording the pedigree of every animal, was kept throughout this phase of the project.

As plans for release experiments developed, it became apparent that a special unit was needed to house animals being prepared for release, and this was set up at Demage Farm on the outskirts of the zoo in 2001.

International Union for the Conservation of Nature (IUCN)

In preparation for the releases it was vital to closely adhere to the protocols issued by the IUCN that govern all mammal reintroductions. The same process has to be carried out whether the plan is to release a tiny mouse in Cheshire, or a major carnivore in Africa. Although it was not the primary aim of the Chester Zoo project, it was important to ensure that if a successful reintroduction resulted from our research, all guidance pertaining to reintroduction had been followed.

Experimental design

In order to write the reintroduction protocol it was necessary to establish a successful release technique that would produce a self-sustaining wild population of harvest mice. After extensive literature searching it was concluded that both soft and hard release techniques were potentially suitable, and that the research should investigate these options. The two main experimental releases were carried out in 2002 and 2003 comparing hard release with two different soft release methods:

- In 2002, using the tank in which they had been previously kept, soft-release animals were placed on the site with a supply of food and water, with the tank open and tipped on its side. Tanks were removed after two days.
- In 2003, soft-release cages were constructed and mice were placed in these for two days with food and water. The cages were then raised from the ground slightly to allow the mice to move in and out as they chose for the next five days. The cages were removed after one week.

Releases were made in early summer (May/June) when food is abundant and harvest mice would normally be breeding. In the 2002 trial, 129 mice were released with a further 270 in the 2003 trial.

Release sites

Release sites were chosen through survey work to establish habitat suitability, existing populations of indigenous small mammals and predator presence. Sites at Chester Zoo were intensively monitored over several years before releases took place, and were found to have large populations of the following small mammals: bank vole, short-tailed field vole, common shrew, pygmy shrew, water shrew and wood mouse. Some management was undertaken to improve these sites before the release trials, which largely took the form of cessation of grazing to allow vegetation cover to improve.

The two sites at Chester Zoo are damp unimproved grassland bordering the Shropshire Union canal, with the site used in 2003 being wetter than that used in 2002. They contain a diverse range of tall grass species, with herbs such as lady's smock, black knapweed and buttercup; reeds and rushes are abundant in the wetter areas. A ditch runs through both sites, which are separated by approximately 500 m of similar habitat.

Release animals

Release animals were carefully selected from individuals between three and 12 months old, in perfect physical condition, with a 1:1 sex ratio.

Every individual underwent a veterinary health assessment, and the group was screened to ensure they were not carrying any diseases or parasites which might affect the animals already living on the release sites. This health screening also extended to the release sites to ensure the resident species were not carrying diseases or parasites at levels that would have a major effect on survival potential and welfare of released harvest mice.

Each mouse destined for release was microchipped, using the smallest available microchip at 8 mm long and weighing 0.3 g. This individual identification has enabled accurate post-release monitoring of animals, helping in the understanding of survival, dispersal and breeding success of the released mice.

In the few weeks prior to release, the diet of the mice was gradually altered to include more wild foods such as grass flowers and seeds, and the proportion of insectivorous food given was increased as it is believed that the diet of wild harvest mice includes a significant amount of insects during summer.

Post-release monitoring

Radio collars were fitted to 20 of the mice released in 2003. This was ground-breaking science as the collars were the smallest ever developed, weighing 0.35 g, and we believe no-one had ever radio tracked an animal as small as a harvest mouse before. Prior to collared animals being released, a small number were observed in captivity for several weeks to ensure that their behaviour and welfare were not affected by the collars. These collared captive animals behaved normally and bred without any observed ill effects.

The batteries on the collars were so small that they only lasted about 20 days so we were able to follow our mice for no more than three weeks from the release date. The 'trackers' worked around the clock following mice and locating them once in each six-hour period. This gave very detailed information about what happened to each individual in the field, specifically survival, dispersal and habitat choice. Both soft- and hard-released animals were tracked and, interestingly, we saw that soft-released animals did return to their protected cages for a few days after release.

Post-release monitoring using small mammal traps commenced 10 days after each release and was repeated at approximately five-weekly intervals throughout the first year until December. After this intense first year, monitoring continued for another three-and-a-half years, focusing on spring and autumn to gauge winter survival and autumn population peaks. Monitoring at each site was done in a consistent manner to ensure that results from different sessions were comparable. Longworth traps and plastic 'trip' traps were used to maximise the efficiency of the process. Trip traps were found to be more successful for harvest mice as they are not only more sensitive, but also more easily set in vegetation above ground level where these mice are active.

Other sites

During 2002 and 2003, the zoo team also supervised the release of a further 150 mice on two more Cheshire sites (Hale Bank, Widnes and Gatewarth near Warrington). These releases followed the same experimental design with both hard and soft releases. In 2004 Cheshire County Council and Biota, a local ecological consultancy, became involved and tested the technique again by releasing 400 mice into Marbury Country Park near Northwich. The data gathered from all of these projects is being pooled to provide as much information as possible.

Results

Analysis of the data is ongoing in 2008 with final monitoring on the zoo sites imminent; scientific publications are also in progress. Initial analysis shows no significant difference in survival between hard- and soft-released animals, or males and females; however, soft-release cages were shown to be used by radio-tracked animals.

Microchipped animals were recaptured soon after release, many in breeding condition, including pregnant females. Some animals travelled a consid-erable distance from the point at which they were released, and individuals have been observed to travel up to 500 m in a day. The first animals without microchips caught at each site were greeted with great excitement as they were evidence of successful breeding.

Importantly, harvest mice have been found on the two zoo sites at almost every trapping session over four years, proving that sustainable populations have been established. Numbers of mice caught in the years after release have fluctuated, as would be expected, with 2003 being the best year. Monitoring on the 2003 zoo site in the spring of 2007 did not look hopeful until the very last day when a single, big healthy female was caught—resulting in a sponta-neous celebration amongst the volunteers!

Bad weather immediately after the release at Marbury Country Park appears to have had a major impact on the animals released there, and it seems likely that the population did not survive.

The project has generated interest at local, national and international levels; a team from Japan even visited Chester Zoo for advice in 2003. Press coverage has been enormous and the crew from the BBC *Springwatch* programme have become good friends with zoo staff. Recollections of filming harvest mice in a wet field with TV naturalist Bill Oddie, and the water shrew that bit him, will live long in the memories of those that were there!

Thanks

It seems likely that some people reading this book will have been involved with this project, or will know people who were. The huge number of dedicated and enthusiastic people who have contributed is one of the main reasons for the success of this project. The project has also provided small mammal trapping training for more than 300 people. There are few sites in Cheshire that can regularly promise seven native small mammal species, but the harvest mouse release sites at Chester Zoo genuinely can.

This project clearly demonstrates Chester Zoo's commitment to British native species conservation, and the zoo's desire to work in partnership with other organisations and individuals to make a significant difference: this is real conservation in action.

Note: Though monitoring for the project ended in 2003, it is likely that small mammal trapping events will continue at Chester Zoo for the foreseeable future as part of routine survey work on the zoo's land. If you require training in small mammal identification or survey techniques please contact the zoo.

14. The future for Cheshire mammals

Introduction

Climate change is arguably the greatest environmental challenge facing the world today, where climate refers to the average weather experienced over a long period including temperature, wind and rainfall patterns. The climate of the Earth is not static: change is an active natural process.

The concern today is that climate change seems to be occurring at a faster rate and that the main driver of this acceleration is human activity: the Earth has warmed by 0.74°C over the last hundred years with around 0.4°C of this warming occurring since the 1970s. Such rises in global temperatures will bring changes in weather patterns, rising sea levels and increased frequency and intensity of extreme weather events, the effects of which will be felt not only here in the UK but all around the world.

The UK currently has a cool, humid and temperate climate, but the effects of cold polar, cool continental and warm tropical air masses, along with oceanic currents and the altitudinal range, mean there is a great spread of climatic conditions and variability in the weather. Greatest mean annual temperatures are recorded in southern Britain, which is on average 3–4°C warmer than in Scotland. The number of wet days varies considerably, as does total annual rainfall.

There is a great deal of uncertainty about how our climate will change in the coming decades, and a number of computer models have been developed to to help us understand climate change. One such is MONARCH (Modelling Natural Resource Responses to Climate Change), a three-phased investigation into the impacts of climate change on the natural conservation resources of the UK and Ireland. This model aims to predict key biodiversity responses to human-induced climate change.

Natural responses to climate change

Individual species have two possible survival mechanisms in response to climate change: to adapt or to move, with the success of each mechanism being dependent upon the severity and rapidity of the change imposed. Evolutionary responses to environmental change are rare as they depend upon there being appropriate variability in the natural population, and often require slow environmental pressures over the long term. Evidence suggests that few vertebrate species will be able to evolve rapidly enough to adapt to the predicted climate changes through natural selection. More likely are shifts in distribution, behavioural changes and possibly extinction.

However, shifts in distribution are constrained. The landscape is formed from a complex patchwork of habitats and land uses that can provide a serious restriction on the capability of even relatively mobile species to move. A species distribution will only change if there is a suitable habitat and food supply along its dispersal route. Range expansion or decline may be observed in species such as the lesser horseshoe bat as it moves northwards. In extreme cases, species that lose their niche may become locally extinct, for example, the mountain hare in Cheshire.

The most obvious effect of climate change will be warmer and potentially drier summers in some parts of the country. An already well-publicised effect of warmer temperatures has been to disrupt normal patterns of hibernation, particularly in relation to hedgehogs. Increasing numbers of animals are becoming active earlier than normal; this can be problematic for species such as dormice as they may have difficulty in finding sufficient food to replenish reserves used over the winter months. By contrast, there is evidence that bats are remaining active virtually throughout the year as insect flight periods become extended, while population numbers are benefitting from the increased availability of insects throughout the spring and summer.

Other possible effects of warmer summer temperatures relate to the habitats in which mammals live. Drought conditions harden the ground reducing the availability of earthworms for badgers, for example. Prolonged periods of dry weather will bring an increased risk of fire leading to habitat loss. There will also be increased pressure on many inland waters, not just through evaporation but also extraction.

Along with an increase in summer temperatures, a similar situation will apply during the winter period. Milder and wetter winters will lead to increased and earlier growth of vegetation during the early spring; this will be particularly favourable to herbivores such as deer and rodents. There will also be an increase in extreme weather events, particularly heavy rainfall over short periods increasing the frequency of flooding. This may not only destroy habitat in general but also possibly affect the breeding success of riparian mammals and water voles in particular. More autumn and winter storms could see increasing losses of mature trees which may have a detrimental effect on woodland species such as bats by reducing the availability of roosting and hibernation sites.

Increasing temperatures may cause expansion of the world's oceans, which, along with meltwater from polar icecaps, will result in a significant rise in sea levels. This will inevitably affect any species that live in coastal habitats such as salt marshes where increased tidal flooding will occur. Any reduction in land area will also be problematic for marine mammals such as seals that need beaches and mudflats to breed and haul out. Disruption in ocean currents will also result, influencing the abundance and distribution of plankton and other marine invertebrates, along with the predators such as fish and cetaceans that feed on them. Whether the increase in sightings of cetaceans in the Irish Sea over recent years is a consequence of this has yet to be determined.

To summarise, the impact of climate change on species and habitats may result in a change in:

- quantity (the abundance of a species, or area of habitat)
- location (the range and distribution of species and habitats)
- quality (the community composition or structure of a habitat, or the genetic variability in a population of a species)

Making predictions about the effect of climate change on mammal populations in Cheshire is an impossible task as there are far too many variables; the occurrence and scope of climate change impacts are unknown.

Extensive data is crucial to help to determine potential adaptation strategies, habitat management and suitable mitigation measures to direct conservation policies and procedures in the future. We must no longer consider habitats in isolation but look towards the wider ecosystem to move forward on how we deal with our response to climate change.

Now is therefore the time to gather as much surveillance and monitoring data as possible, both on a local level and on a wider scale. Surveys need to look further than traditional methodologies and need to include the measurement of other variables such as temperature, humidity and rainfall.

A call to arms

An appreciation and understanding of our natural environment has been a necessity ever since our ancestors first walked upright. Knowing and understanding the world around them meant the continued survival of our species by finding adequate food, water and shelter, as well as avoiding predators. As time went on, the potential for humans to influence the natural environment was demonstrated through movement from a nomadic, hunter-gatherer existence to a settled agricultural way of life. Understanding the natural environment became even more important, especially for food production.

Much of this knowledge was passed down from generation to generation through oral tradition. In recent centuries a mass of knowledge and experience has been transferred into written form, to ensure retention of this invaluable resource. However, this written wisdom only guides us, it can never replace up-to-date knowledge gained from first-hand experience of nature with an expert.

In the modern world, and certainly in the UK, the majority of the population is focused around large cities with only limited exposure to the natural environment. Many people would not know, for example, that the number of small mammals can decline after a run of wet springs. It may be said that most of us don't need to know such things, because there are people out there who monitor these species. To a certain extent this is true. However, today there are not enough people with even a passing interest in recording wildlife, let alone enough with an adequate level of expertise to monitor the situation.

Many of the best naturalists in the UK today are of the silver-haired generation. They are battling to monitor our environment to ensure that what we have now is recorded and retained for future generations, and to pass on their knowledge. Is it right that they should carry this burden on their own? This is where local wildlife groups come in with organisations such as Cheshire Mammal Group who welcome and train new members to assist with recording the wildlife of the county. This work has a vital part to play not only in planning for the environmental effects of challenges such as climate change, but also in working to protect habitats and species for future generations, thus ensuring the continuation of our green and pleasant land.

For information on how you can get involved with Cheshire Mammal Group visit the rECOrd website (www.rECOrd-LRC.co.uk) and look for the Cheshire Mammal Group page. This gives details of training events and meetings so you too can get involved and see at first hand how fascinating the mammals of Cheshire are!

Appendix 1. Cheshire region Mammal Action Plans

Bats

The action plan applies to the nine bat species that have recently been recorded in the Cheshire region namely: common and soprano pipistrelle (*Pipistrellus pipistrellus* and *Pipistrellus pygmaeus*), brown long-eared (*Plecotus auritus*), whiskered (*Myotis mystacinus*), Brandt's (*Myotis brandtii*), noctule (*Nyctalus noctula*), Daubenton's (*Myotis daubentonii*), Leisler's (*Nyctalus leisleri*) and Natterer's (*Myotis nattereri*). The soprano pipistrelle, noctule and brown long-eared are the only Cheshire region bats included on the UK BAP priority list.

Species	Maintenance target	Expansion target
Pipistrellus pipistrellus	117	129 (10%)
Pipistrellus pygmaeus	62	71 (15%)
Plecotus auritus	66	73 (10%)
Myotis mystacinus	4	12 (300%)
Myotis brandtii	2	3 (50%)
Nyctalus noctula	65	72 (10%)
Myotis daubentonii	35	39 (10%)
Nyctalus leisleri	1	2 (100%)
Myotis nattereri	8	10 (20%)

National targets

Soprano pipistrelle

Target 1: Maintain soprano pipistrelle population above 2005 population index baseline level.
Target 2: Increase soprano pipistrelle population index by 35 per cent of the 2005 baseline level by 2020.

Notes: Target 1 is based on maintaining the population index calculated from National Bat Monitoring Programme (NBMP) field surveys at 100 or above (i.e. there is no decrease in the population index indicative of a decline in actual population).

Other species

Noctule and brown long-eared bats were added to the UK BAP priority list in June 2007. National targets for these species are awaited. Common pipistrelle has been removed from the UK BAP priority list.

Local targets

Target 1: Maintain the current range of all bat species at the 2007 baseline level of occupied tetrads in the Cheshire region.
Target 2: Achieve an increase in the range of each bat species in the Cheshire region by 2015.

Priority actions

Recording and monitoring

- Monitor known population and record and validate new sightings using volunteers from the Cheshire Bat Group and the Merseyside and West Lancashire Bat Group as part of the NBMP.
- Collate records from bat workers, members of the public and other sources and submit to the local biological records centre.

Advisory

- Protect threatened roosts and individual bats through a system of consultation between Natural England and the local bat groups.
- Influence development mitigation to include appropriate building and landscape features for bats.

Site safeguard and management

- Influence Environmental Stewardship agreements to include options beneficial to bats.
- Work in partnership with wetland, woodland and hedgerow habitat action plans to influence best practice habitat management for bats.

- Install and monitor artificial roosts at key sites across the region.
- Encourage landowners to enhance roosting and foraging habitat through sensitive farming practices towards restoration of a traditional landscape and boundary features.
- Encourage maintenance and enhancement of wetlands and waterbodies to increase the availability of insect prey.

Communications and publicity

- Support trainee bat workers to gain the experience required to obtain a bat license.
- Raise public awareness of bats and their protective legislation through educational activities, training events and media coverage.
- Licensed bat workers will advise on and promote responsible roost ownership.

Brown hare   *Lepus europaeus*

National targets

Target 1: Maintain current range of brown hare at 1,604 occupied 10 km squares in the UK.
Target 2: Increase the population size of brown hare to double the 1995 level (population index = 100) by 2010.

Units	2005 baseline		2010 target		2015 target	
	UK	Eng	UK	Eng	UK	Eng
Population index	150	150	200	200	200	200

Notes: The general aim is to retain the brown hare as a common farmland animal. Target 1 requires the maintenance of the current distribution so that there is no net loss of occupied 10 km squares whilst target 2 intends to restore numbers to half of what they had been at the beginning of the twentieth century.

Local targets

Target 1: Maintain the current range of brown hare at 338 occupied tetrads in the Cheshire region.
Target 2: Achieve a 10% increase in occupied tetrads, to 372, by 2015

Priority actions

Recording and monitoring

- Carry out annual spring vantage point and transect surveys at selected sites in the region to monitor population trends.
- Collate local records as part of the National Hare Survey and submit to the local records centre and Joint Nature Conservation Council.
- Update the Cheshire region-wide survey completed in 2000.

- Map core hare population 'hot spots' to identify key areas for protection and habitat improvements.
- Monitor and record incidence of poaching.

Advisory

- Encourage 'hare-friendly' farming practices.
- Distribute management advisory guidance to landowners.

Site safeguard and management

- Consider the requirements of brown hare in the development of Environmental Stewardship agreements.
- Reduce fragmentation of the brown hare population through targeted habitat improvements.
- Encourage the uptake of set-aside schemes and stubble strips to provide cover and food for brown hares.

Communications and publicity

- Use the popularity of brown hares to raise the profile of biodiversity and the impacts of intensive farming upon local wildlife.
- Train volunteers in hare survey methodology.
- Promote recording of mammals by the general public by using brown hare as a flagship species.

Hazel dormouse *Muscardinus avellanarius*

National targets

Target 1: Maintain the current range of dormouse at 376 occupied 10 km squares in the UK (excluding reintroduced populations).
Target 2: Re-establish self-sustaining dormouse populations at 16 sites, in counties where they have been lost, by 2010.
Target 3: Ensure the dormouse population index is at 100 per cent of the 1991 level by 2015 and increase to 115 per cent of the 1991 level by 2020.

Units	2005 baseline		2010 target		2015 target	
	UK	Eng	UK	Eng	UK	Eng
Sites	11	11	16	16	21	21
Population index	Trend –ive	Trend –ive	Trend 0/+ive	Trend 0/+ive	100	

Notes: Target 3 involves converting the 2005 negative population trend into a stable or positive population trend by 2010 to restore declining dormice populations to the 1991 level.

Local targets

Target 1: Maintain the current range of the reintroduced dormouse population at two occupied tetrads in the Cheshire region.
Target 2: Achieve an increase in the range of the dormouse from two occupied tetrads in 2007 to three by 2015.
Target 3: Increase the pre-breeding dormouse population index within the Wych Valley from 100 in 2007 to 150 in 2015.
Target 4: Identify and complete pre-survey for one further reintroduction site in the Cheshire region by 2015.

Notes: Target 3 is based on the pre-breeding adult dormouse catch in 2007 of 22 (population index = 100). Target 3 aims to increase the adult dormouse catch to at least 33 (population index = 150) by 2015.

Priority actions

Recording and monitoring

- Monitor nest-box usage four times annually in the Wych Valley as a key site in the National Dormouse Monitoring Programme.
- Monitor and record breeding success and patterns of dispersal.
- Use nest tubes to monitor dormouse activity at historic sites not previously surveyed and potential reintroduction sites.
- Carry out population studies using microchipping and DNA analysis.

Advisory

- Research dormouse ecology to inform the production of woodland management guidelines.
- Consider landscape scale and regional actions for dormice by forging links with habitat action groups and neighbouring county dormouse groups.
- Influence woodland and hedgerow management across the region to establish a wide network of suitable habitat.
- Work cross-border with Welsh agencies to encourage a partnership approach to monitoring.

Site safeguard and management

- Encourage population growth and dispersal by securing appropriate long-term management for the Wych Valley and immediate surroundings.
- Survey woodlands and hedgerows in the vicinity of the current site to identify areas for colonisation.
- Identify suitable sites for future reintroductions and secure appropriate management agreements with landowners.

Communication and publicity

- Recruit and train dormouse volunteers through practical days and workshops.
- Encourage public participation through the 'Great Nut Hunt'.
- Publicise the Cheshire Dormouse Project through newsletter and magazine articles.
- Promote the Cheshire Dormouse Project as a flagship biodiversity conservation project in the region.

European otter *Lutra lutra*

National targets

Target 1: Maintain the current distribution of the otter throughout the UK.
Target 2: Expand the distribution of otters to achieve 85 per cent occupancy of 10 km squares by 2015.

Units	2005 baseline		2010 target		2015 target	
	UK	Eng	UK	Eng	UK	Eng
Occupied 10 km squares	2219	878	2377	997	2500	1084

Notes: Current baseline data for target 1 indicates that 74.2 per cent of 10 km squares in the UK are occupied by otter. The majority of the range expansion in target 2 is expected to take place in England and will occur through natural recolonisation once threats to otter survival have been removed.

Local targets

Target 1: Maintain the current distribution of the otter throughout the Cheshire region at 47 occupied tetrads.
Target 2: Achieve an increase in the range of otters from 47 occupied tetrads in 2007 to 90 occupied tetrads by 2015.

Priority actions

Recording and monitoring

- Keep the database of records in Cheshire updated through repeat surveys of sites in line with the seven-year rolling programme of national surveys.
- Visits to sites included in the national survey will be supplemented by more local surveys and monitoring to inform the assessment of full distribution.
- The recovery of otters and their return to currently unoccupied reaches will be closely monitored.
- Record and monitor threats to otter survival. In particular road casualties will be collected for DNA analysis and their location used to inform highways mitigation projects.

Advisory

- Ensure otters are taken into consideration in the planning process. This will include providing training and advice for local authority planners and highways engineers.
- Ensure that all new bridges include potential spranting sites such as dry ledges and encourage the installation of otter-friendly features to all existing bridges over 6 m wide.

Site safeguard and management

- Monitor the condition and usage of existing artificial and known natural holts.
- Encourage landowners and developers to install new artificial holts at appropriate sites.
- Encourage population expansion by natural colonisation of former sites through habitat enhancement. This will include advising landowners on sympathetic riparian habitat management and encouraging the uptake of options beneficial to otters in Higher Level Stewardship schemes.
- Work in partnership with statutory agencies to improve and then maintain good water quality across all catchments in the Cheshire region.
- Work in partnership with wetland habitat action plans to secure the expansion of suitable floodplain habitat.

Communication and publicity

- Raise awareness and understanding of the threats to otters through training events and promotional activities.
- Recruit and train volunteer surveyors.
- Produce and distribute riverside management advice to local landowners.
- Use the popularity of otters to promote issues affecting the biodiversity of riparian systems.

Water vole *Arvicola terrestris*

National targets

Target 1: Maintain the current range (i.e. no net loss) of water voles at 730 occupied 10 km squares in the UK.

Target 2: Achieve an increase in the range by 50 new occupied 10 km squares in the UK by 2010.

Units	2005 baseline		2010 target		2015 target		2020 target	
	UK	Eng	UK	Eng	UK	Eng	UK	Eng
Occupied 10 km squares	730	582	780	605	835	635	895	675

Local targets

Target 1: Maintain the range of water voles at the 2007 baseline of 195 occupied tetrads.

Target 2: Achieve an increase in the range of water voles from 195 occupied tetrads in 2007 to 234 by 2015.

Priority actions

Recording and monitoring

- Establish the current distribution and range of water voles in the Cheshire region. Compile and keep up-to-date baseline survey information on the status of water voles across the region.
- Monitor the status of water voles annually at sites identified as part of the National Water Vole Survey 1997/98. Supplement this with county-wide surveys at least once every five years.
- Identify trends in populations linking data where possible to metapopulation dynamics and changes in habitat condition and/or threats.
- Establish the current distribution and monitor the effects of mink at key water vole sites.

Advisory

- Ensure that development schemes do not affect the integrity of water vole populations through the development control process.
- Impose site safeguard and suitable mitigation measures as conditions of developments.
- Consider water voles in the designation of local and statutory sites.
- Seek inclusion of water vole habitat management in riparian management plans including Local Environment Agency Plans.

Site safeguard and management

- Safeguard water vole populations against mink through active control in partnership with landowners and landowner associations.
- Promote favourable management of riparian habitats to maintain the current distribution. This includes targeting sites where water vole populations have been recently lost for habitat improvements and/or predator control. A twin-track approach of proactive habitat and mink management will be used to allow natural range expansion into adjacent currently unoccupied 10 km squares.
- Promote Environmental Stewardship options that benefit water voles and wetland habitats.
- Promote best practice habitat management in line with the Water Vole Conservation Handbook.

Publicity and communication

- Increase awareness of water voles and their conservation issues. This includes providing interpretation at publicly accessible sites and media coverage of water vole issues.
- Organise survey training events to recruit volunteers and support their skill development.
- Work with anglers and other water users to coordinate conservation efforts and records of sightings.

Small cetaceans

This action plan applies to five small cetaceans species which occur regularly in the Irish Sea off the Wirral coast namely harbour porpoise (*Phocoena phocoena*), bottlenose dolphin (*Tursiops truncatus*), Risso's dolphin (*Grampus griseus*), white-beaked dolphin (*Lagenorhynchus albirostris*) and common dolphin (*Delphinus delphis*). Following the 2007 UK BAP review, all five species are now classified as UK priorities.

National targets

Harbour porpoise

The harbour porpoise has been a UK BAP priority species since 1994. However, national targets are still awaited.

Other species

The bottlenose dolphin, Risso's dolphin, white-beaked dolphin and common dolphin have recently been recognised as nationally important and were added to the UK BAP priority list in June 2007. National targets for these species have yet to be published.

Local targets

Target 1: Maintain the range of harbour porpoise at the 2007 baseline level of 7 occupied tetrads.
Target 2: Raise awareness, understanding and monitoring of small cetaceans by delivering at least five events each year.
Target 3: Establish a new project in partnership with Liverpool Bay to encourage monitoring of small cetaceans.

Priority actions

Recording and monitoring

- Establish a baseline study of population and conservation status of small cetaceans in Liverpool Bay and the Dee Estuary.
- Monitor small cetacean populations and threats to their survival by utilising local volunteers and Coastal Networks.
- Maintain a database of small cetacean sightings.
- Collect tissue samples from corpses to enable post-mortem analysis.

Advisory

- Promote the implementation of DEFRA guidelines to reduce acoustic and recreational disturbance of small cetaceans.
- Encourage a partnership approach to coastal and marine conservation.
- Liaise with fisheries to promote best practice methods that reduce the incidence of accidental by-catch of small cetaceans.

Site safeguard and management

- Influence local policy and codes of conduct to improve coastal water quality.
- Represent the needs of small cetaceans where planning developments affect estuary ecosystems.
- Research the population status and conservation needs of small cetaceans around Cheshire to inform management options to safeguard their future.
- Lobby for better protection of the marine environment.

Publicity and communication

- Encourage recording of sighting and disturbance events by local people.
- Promote the presence of small cetaceans through marine-themed events and site interpretation.

White-beaked dolphin.

DAVID QUINN

Grey seal *Halichoerus grypus*

National targets

There are no national targets for this species.

Local targets

Target 1: Maintain the range of Atlantic grey seals at the 2007 baseline of eight occupied tetrads.
Target 2: Increase the public and environmental professional's understanding of grey seals particularly amongst those working on or using estuaries and coastline for leisure and recreation by delivering at least five events each year.

Priority actions

Recording and monitoring

- Continue counting seals daily (whenever personnel are available), log information at Hilbre Bird Observatory and pass information to rECOrd.
- Develop guidance and protocols for monitoring which can be used throughout the Irish Sea.
- Encourage research projects about the seals through liaison with local universities.

Advisory

- Compile and maintain a database of Dee and Mersey Estuary recreational users. Advise these users of codes of conduct that minimise disturbance to grey seals.
- Contact those known to fish in the area and advise of best practice methods to avoid disturbance and accidental death of grey seals.
- Compile appropriate leaflets and posters including a code of conduct for wildlife boat trips, to promote the reduction of disturbance to seals.

Site safeguard and management

- Consider the usefulness of and introduce, where necessary, zonation to protect seal populations.
- Monitor any commercial activity that would affect the grey seal population.

Publicity and communication

- Maintain and update list of interested parties including recreational users and fisheries and circulate annually with up-to-date information.
- Annually circulate appropriate leaflets and posters including a code of conduct for wildlife boat trips.
- Continue providing an information service to schools, other groups and general visitors to Hilbre.
- Promote awareness and understanding of the grey seal population.

Harvest mouse *Micromys minutus*

National targets

The harvest mouse has been recognised as of national importance and was added to the UK BAP priority list in June 2007. National targets for this species have yet to be published.

Local targets

Target 1: Establish the range and status of wild harvest mouse populations in the Cheshire region by 2015.
Target 2: Publish a written protocol for use in the reintroduction of harvest mice around the world where and when needed by 2010.
Target 3: Maintain an active annual program of post-release monitoring to determine the survival rate and dispersal of introduced populations.

Priority actions

Recording and monitoring

- Encourage further recording of this species across the Cheshire region to establish the baseline distribution of wild harvest mice. Carry out surveys using both small mammal trapping and spring nest search methods.
- Ensure accurate species identification and verify those records already in existence where this has not already been achieved.
- Monitor the status of harvest mice at sites where harvest mice have been recorded historically and where captive-bred individuals have been released.
- Use post-release monitoring to further the reintroduction protocol.

Site safeguard and management

- Work with universities to carry out research that determines habitat preferences and usage.
- Record and compare vegetation at a range of sites with wild and reintroduced populations to further determine habitat choice.
- Encourage uptake of options that benefit harvest mice in Environmental Stewardship schemes.

Publicity and communication

- Compile and publish the results of the captive breeding and release programme.

Polecat *Mustela putorius*

National targets

The polecat has been recognised as of national importance and was added to the UK BAP priority list in June 2007. Nationals target for this species have yet to be published.

Local targets

Target 1: Maintain the range of polecats at the 2007 baseline level of 69 occupied tetrads.
Target 2: Achieve an increase in the range of polecats for 69 occupied tetrads in 2007 to 91 by 2015.

Priority actions

Recording and monitoring

- Promote and encourage recording of polecats across the Cheshire region, particularly amongst the farming community.
- Collate distribution data and monitor the status of polecats. Submit all sightings to the local records centre.
- Consider the use of live trapping and radio tracking to further the understanding of polecat ecology.

Site safeguard and management

- Research the habitat usage and preferences of polecats to influence land management regimes.

Publicity and communication

- Liaise with farmers and landowners on the issues and threats facing polecats and the steps needed to conserve them.
- Produce interpretive material and media coverage to increase local understanding of polecats.

Appendix 2. Mammal legislation

Species	UK	European	UK BAP	LBAP
Common shrew	1			
Pygmy shrew	1			
Water shrew	1			
Bats	2	8, 9	✓	✓
Brown hare	4, 5		✓	✓
Mountain hare		10		
Red squirrel	2		✓	
Water vole	2, 3		✓	✓
Harvest mouse				✓
Common or hazel dormouse	2	13	✓	✓
Harbour or common porpoise	2	8, 9, 10, 11, 12		✓
Short-beaked common dolphin	2	8, 9, 10, 11, 12		✓
Northern bottlenose dolphin	2	8, 9, 10, 11, 12		✓
White-beaked dolphin	2	8, 9, 10, 11, 12		✓
Orca	2	8, 9, 10, 11, 12		✓
Common seal	6			
Grey seal	6	8, 10	✓	✓
Pine marten	1			
Polecat	1			✓
Badger	7			
Otter	2, 3	8, 10, 12	✓	✓

1. Wildlife and Countryside Act (1981) as amended. Schedule 6
2. Wildlife and Countryside Act (1981) as amended. Schedule 5
3. Wildlife and Countryside Act (1981) as amended. Schedule 8
4. Ground Game Act (1880)
5. Hare Protection Act (1911)
6. Conservation of Seals Act (1970)
7. Protection of Badgers Act (1992)
8. The Conservation (Natural Habitats etc.) Regulations 1994—Annex II
9. The Conservation (Natural Habitats etc.) Regulations 1994—Annex IV
10. The Conservation (Natural Habitats etc.) Regulations 1994—Annex V
11. Bonn Convention Appendix II
12. Bern Convention Appendix II
13. Bonn Convention Appendix III

Appendix 3. Local groups

Cheshire Mammal Group (CMaG)

Established in 2001, the Cheshire Mammal Group aims to encourage the study, conservation and awareness of mammal species in the Cheshire region.

The aims of the group are to provide information and practical help for national and local mammal surveys, to support and promote the work of the Mammal Society and to interest new people in mammals.

The group is involved with various activities including:

- Training sessions in mammal identification and survey methods
- Recording visits held throughout the year
- Public education events about local mammals: talks, walks and practical sessions

The group is affiliated to the national Mammal Society and works closely with both the Wirral and Cheshire Badger Group and the Cheshire Bat Group.

For information on how you can get involved with CMaG, visit the rECOrd website and look for the CMaG page. The group meets on a three-monthly basis: if you wish to come along to a meeting please contact us.

Website: www.rECOrd-LRC.co.uk

Email: info@rECOrd-LRC.co.uk

Address: c/o rECOrd, Oakfield House, Chester Zoological Gardens, Upton, Chester, CH2 1LH

Tel: 01244 383749

Cheshire Bat Group

The Cheshire Bat Group has been established since 1986 with enthusiastic, active and knowledgeable members who share an interest in bats and bat conservation. This interest takes many forms and members are involved in

- Survey work of bat distribution, roosts, hibernacula and feeding areas
- Bat promotional work through organised walks and talks
- Home visits to householders needing advice and information about bats
- Bat care and rehabilitation
- Research and data collection and collation
- Training sessions
- Practical bat conservation activities

In recent years the Cheshire Bat Group has been actively involved in the Cheshire Bat BAP, alongside Cheshire Wildlife Trust, to further the aims of conservation and understanding in the Cheshire region.

The group covers the administrative areas of Cheshire, Halton, Warrington, Stockport, South Manchester, Altrincham and Hale and has links to adjoining County bat groups. Members do not have to live in Cheshire.

Cheshire Bat Group is keen to learn more about all of the species occurring in the county, and conservation and data-gathering efforts target all species.

Email: cheshirebats@yahoo.co.uk

For grounded or injured bats or problems with bats in buildings contact the Bat Conservation Trust who will be able to directly contact a local volunteer.

Bat Conservation Trust helpline, tel: 0845 1300 228

Chester Zoo

The mission of Chester Zoo is 'to be a major force in conserving biodiversity worldwide,' and working here in the UK is considered a very important part of that mission. Chester Zoo has always taken local conservation seriously as demonstrated by long-term involvement with species such as harvest mice, sand lizards and barn owls. The zoo's contribution to UK conservation has grown significantly since 2000 with the appointment of a Biodiversity Officer, dedicated to finding useful roles for the zoo in UK conservation, and engaging with partners and projects.

Chester Zoo can provide a wide range of skills and facilities, and the contributions the zoo can make to conservation are diverse. Amongst the staff there is wide-ranging scientific knowledge, many individuals have detailed specific expertise and experience, and survey skills and husbandry techniques applicable to a wide range of native species are available. In addition the zoo has horticultural expertise and practical skills in landscaping and habitat creation, and a highly skilled veterinary team. Chester Zoo staff have lots of enthusiasm for conservation projects too.

Chester Zoo also offers *ex situ* care facilities, meeting rooms, horticultural services, a veterinary laboratory, training courses and survey equipment such as small mammal traps and radio-tracking gear.

It should not be overlooked that the zoo welcomes over a million visitors per year and enjoys the support of more than 30,000 members. Chester Zoo is one of the most popular wildlife attractions in the UK and provides a unique opportunity to pass on conservation messages to the public and to motivate people about conservation issues.

Chester Zoo supports conservation work at a range of different levels:

- Conservation programmes: based on strong partnerships with substantial financial and practical zoo commitment guaranteed for several years. Focus may be on a region, habitat or species, with multiple project components and an emphasis on long-term sustainability. There are currently eight programmes, one of these being the UK Native Species Programme.
- Conservation and research grants: one-off funding support provided for projects, chosen by a review process to select those which best complement the zoo's conservation mission and existing activities.
- Studentships and scholarships: available to students for very varied projects. Applications are judged as much on the future potential of the student to contribute to conservation, as on the quality of the project.
- 'Keeper for a day' activities: funds obtained from visitors paying to come and work at Chester Zoo for a day are used to enable zoo staff to contribute to conservation projects across the world.
- Technical support: for example, advice from Chester Zoo bird keepers on hand-rearing techniques to *in situ* breeding projects abroad, and advice from curators and vets on handling and monitoring techniques.
- Fundraising campaigns: through promotional events and advertising etc.

The UK Native Species Programme

Chester Zoo's contribution to UK conservation projects includes regional coordination through involvement in the LBAP process in north-west England and North Wales, and the facilitation of partnerships locally and across the UK. At project level, zoo roles vary from lead partner to practical or advisory input, and funding support. The list of native species projects in

which the zoo plays a part is constantly evolving, but an idea of the variety is provided by the list below.

Taxon-based projects

- Harvest mouse reintroduction research
- Dormouse research in Cheshire and North Wales
- Brown hare distribution study in North Wales
- Barn owl survey and nest-box provision in Cheshire and North Wales
- Sand lizard *ex situ* breeding and reintroduction work
- Freshwater pearl mussel conservation in the Dee
- Bat survey work at the zoo
- Black poplar propagation work at Chester Zoo
- MacKay's horsetail insurance population held at the zoo
- Isle of Man cabbage insurance population held at the zoo
- Limestone woundwort propagation and reintroduction work in North Wales
- Juniper propagation and reintroduction work in North Wales
- Welsh cotoneaster propagation research

Habitat-based projects

- Gardens in Cheshire: promoting wildlife-friendly gardening practices
- Pond surveys in Wrexham

For more information on Chester Zoo's work with UK species contact the conservation department: conservation@chesterzoo.org

Website: www.chesterzoo.org

Address: Chester Zoo, Cedar House, Caughall Road, Upton-by-Chester, Chester CH2 1LH

Tel: 01244 380280

Wirral and Cheshire Badger Group

The Wirral and Cheshire Badger Group is a voluntary organisation which helps protect badgers in and around the region and, through our links with the Badger Trust, across the UK. The group was formed in 1980 to combat the regular digging, torturing and killing of badgers in Cheshire and Wirral. At this time the number of badger setts had been reduced by a massive 80 per cent in 10 years, mainly due to persecution. In those days methods were to arrange patrols of vulnerable setts and, through talks, displays, radio and television appearances and road shows, encourage people to get involved with the protection of badgers and to report any suspicious activity to the police.

Progress has been made since then. Badger persecution has dramatically reduced in recent years because penalties including fines and jail terms have increased and the equipment used by badger diggers is often confiscated—vehicles and dogs included. But persecution still goes on ...

Nowadays the activities of the group are more about education and raising of awareness. They spread the word through talks, country shows and badger-watching evenings, and continue to offer advice and support to anyone with badger-related issues.

Website: www.wcbg.org.uk

Address: Wirral and Cheshire Badger Group, PO Box
          19, Warrington WA2 8TG

Tel: Guy Lingford, 01270 582985

Cheshire Wildlife Trust

Cheshire Wildlife Trust is a charity that works to protect and enhance wildlife in the Cheshire region. As part of their role to safeguard the county's fragile natural heritage, they manage 45 nature reserves which help to protect endangered species, rare plants and threatened habitats. They are all about people taking action for wildlife at a local level.

As the natural environment including ponds, woodland, grasslands and peatlands is disappearing, and wildlife heritage is being lost to development, pollution and intensive farming, it is all the more important to preserve the unique local area and its wildlife to help maintain the balance of nature. Therefore the Trust are taking on a broader role within the region in the light of the increasingly important environmental agenda.

The work of Cheshire Wildlife Trust is carried out by a wide range of people: there is a small permanent staff at their headquarters, Bickley Hall Farm, as well as a few volunteers. Staff are organised into teams: Estates and Land Management (looking after the reserves), Conservation (looking after biodiversity in the wider county), People and Wildlife (WATCH groups and wildlife education) and Business and Development (looking after the members).

There are nine local groups each with their own chairman and committee of officers. Each group has their own programme of meetings and events which are ideal opportunities to meet other local wildlife enthusiasts. Most local group meetings are open to non-members. Some of the members are committed and dedicated naturalists whilst others are simply people who enjoy the Cheshire countryside and its wildlife.

Website: www.cheshirewildlifetrust.co.uk

Email: info@cheshirewt.cix.co.uk

Address: Cheshire Wildlife Trust, Bickley Hall Farm,
          Bickley, Malpas, Cheshire SY14 8EF

Tel: 01948 820728

Appendix 4. National and international mammal organisations

The Badger Trust

Website: www.nfbg.org.uk
Address: PO Box 708, East Grinstead RH19 2WN
Tel: 08458 287878
Fax: 02380 233896

Bat Conservation Trust

Website: www.bats.org.uk
Address: 15 Cloisters House, 8 Battersea Park Road, London SW8 4BG
Tel: 020 7627 2629
Bat Helpline: 0845 1300 228

The British Deer Society

Website: www.bds.org.uk
Address: The Walled Garden, Burgate Manor, Fordingbridge, Hampshire SP6 1EF
Tel: 01425 655434

British Hedgehog Preservation Society

Website: www.britishhedgehogs.org.uk
Address: Hedgehog House, Dhustone, Ludlow, Shropshire SY8 3PL
Tel: 01584 890801

The Mammal Society

Website: www.abdn.ac.uk/mammal
Address: 3 The Carronades, New Road, Southampton SO14 0AA
Tel: 02380 237874

Peoples Trust for Endangered Species
(now incorporating Mammals Trust UK)

Website: www.ptes.org
Address: 15 Cloisters House, 8 Battersea Park Road, London SW8 4BG
Tel: 020 7498 4533

Sea Mammal Research Unit

Website: www.smru.st-and.ac.uk
Address: Gatty Marine Laboratory, University of St Andrews, St Andrews, Fife KY16 8LB
Tel: 0 1334 462630

Sea Watch Foundation

Website: www.seawatchfoundation.org.uk
Address: PO Box 3688, Chalfont St Peter, Gerrards Cross SL9 9WE
Tel: 0845 202 3892

Tracking Mammals Partnership (Joint Nature Conservancy Council)

Website: www.jncc.gov.uk

Whale and Dolphin Conservation Society

Website: www.wdcs.org.uk
Address: Brookfield House, 38 St Paul Street, Chippenham, Wiltshire, SN15 1LJ
Tel: 01249 449500

WWF
(formerly Worldwide Fund for Nature)

Website: www.wwf.org.uk
Address: Panda House, Weyside Park, Godalming, Surrey GU7 1XR
Tel: 01483 426444

Appendix 5. Gazetteer

Place	Grid ref
Abbots Moss	SJ595688
Acton Bridge	SJ601760
Adlington	SJ913802
Alderley Edge	SJ843787
Appleton	SJ615840
Audlem	SJ660438
Beeston Castle	SJ538593
Birkenhead	SJ320890
Blakemere	SJ549709
Bolesworth	SJ493560
Bosley	SJ91–67–
Bosley Minn	SJ935660
Bowdon	SJ75–86–
Broadbottom	SJ994938
Bromborough	SJ350827
Budworth Mere	SJ655769
Bunbury	SJ568578
Burland	SJ615535
Chelford	SJ82–73–
Chester	SJ406665
Chester Zoo	SJ413704
Combermere	SJ583443
Congleton	SJ860630
Coppenhall	SJ69–56–
Cranage	SJ754685
Croughton	SJ41–72–
Cuddington	SJ44–54–
Delamere	SJ540710
Dunham Park	SJ735873
Eastham	SJ360800
Egremont	SJ319920
Ellesmere Port	SJ400770
Farndon	SJ415545
Fermilee Reservoir	SK015768
Frodsham	SJ520782
Gatewarth	SJ57–87–
Gayton	SJ24–87–
Griffiths Road Lagoons	SJ686735
Haddocks Wood	SJ545829
Hale Bank	SJ485835

Place	Grid ref
Hartford	SJ64–71–
Hatchmere	SJ553722
Helsby	SJ490757
Henbury	SJ856731
Hilbre	SJ185883
Holcroft Moss	SJ680933
Hoylake	SJ218894
Kingsley	SJ550746
Knutsford	SJ753785
Leasowe	SJ270915
Lindow Common Local Nature Reserve	SJ834810
Little Budworth	SJ585655
Longdendale	SK053984
Lyme Park	SJ970830
Lymm	SJ680875
Macclesfield	SJ920735
Malpas	SJ488472
Marbury Country Park	SJ653763
Marton	SJ68–65–
Mere Hall	SJ843766
Middlewich	SJ704663
Milton Green	SJ462588
Mobberley	SJ784798
Moore Nature Reserve	SJ59–86–
Nantwich	SJ652523
Neston	SJ292775
Nether Alderley	SJ843766
New Brighton	SJ305935
Northenden	SJ828901
Northwich	SJ656736
Norton	SJ555842
Oakmere	SJ574678
Parkgate	SJ275788
Petty Pool SSSI	SJ618701
Pickering's Pasture Local Nature Reserve	SJ488835
Pick mere	SJ683772
Poynton	SJ920836
Prestbury	SJ903775
Rainow	SJ951761
Risley Moss	SJ660915

Place	Grid ref	Place	Grid ref
Rostherne	SJ745852	Tytherington	SJ910760
Rudheath	SJ743708	Wallasey	SJ310935
Runcorn	SJ515820	Warrington	SJ600880
Sandbach	SJ760609	West Kirby	SJ212861
Shavington	SJ699515	Widnes	SJ513853
Stockport	SJ91–88–	Wildboarclough	SJ983689
Swettenham	SJ801673	Wilmslow	SJ845810
Tarporley	SJ554625	Wistaston	SJ679535
Tatton Park	SJ750815	Woolston	SJ640895
Thornton-le-Moors	SJ442745	Wrenbury	SJ594476
Thurstaston	SJ248842	Wybunbury Moss	SJ698502
Trentabank Reservoir	SJ963713	Wych Valley	SJ46–44–

Glossary

aggregation	A society or group of organisms which has a social structure
arboreal	Adapted for living in trees
BAP	Biodiversity Action Plan
biodiversity	The number and variety of plant and animal species that exist in a particular environmental area or in the world generally
caudal patch	A patch of hairs on the rump of a deer
cephalopod	A carnivorous mollusc from the group which includes squid, cuttlefish and octopus
crustaceans	Primitive animals characterised by two pairs of antennae, found in marine, freshwater and terrestrial habitats
cusped	With pointed projections (teeth)
DEFRA	Department for Environment, Food and Rural Affairs
dorsal	On or near the back of an animal
ecosystem	A community of organisms and their physical environment
falcate	Sickle shaped
gastropod	A mollusc from the group that includes snails
heterodyne	A method of combining frequencies to produce a new frequency equal to the sum or the difference between the two
hydrosere	The natural zonation of vegetation at the edges of freshwater habitats.
LBAP	Local Biodiversity Action Plan
Longworth trap	A widely used humane small mammal trap
pedicle	The base from which the antler develops in deer
pelage	The fur of an animal
prism	A polyhedron with two parallel opposite faces
rECOrd	The local biological records centre for Cheshire
riparian	Living or situated on the banks of rivers or streams
tetrad	A 2 km × 2 km square on the Ordnance Survey National Grid System
tibia	The bone which can be felt at the front of the lower leg
tines	The spikes or prongs of an antler in deer
torpor	A dormant state
tragus	A lobe developed from the lower rim of the ear in bats
zygote	Cell formed by the fertilisation of an egg by sperm

Further reading

Aitkenhead, N. *et al.*, *The Pennines and adjacent areas, British Regional Geology Guide* (Nottingham: British Geological Survey, 2002).

Altringham, J.D., *British Bats (New Naturalist Series)* (London: Collins, 2003).

Arnold, H.R., *Atlas of mammals in Britain* (London: HMSO, 1993).

Barratt, E.M., Deaville, R., Burland, T.M., Bruford M.W., Jones, G., Racey P.A. & Wayne, R.K., 'DNA answers the call of the pipistrelle bat species', *Nature*, 387 (1997), 138–39.

BBC website: www.bbc.co.uk

Biodiversity Reporting and Information Group, *Report on the Species and Habitat Review* (Peterborough: Joint Nature Conservation Committee, 2007).

Birks, J., *The Pine Marten* (London: The Mammal Society, 2003).

Bouchardy, C. & Moutou, F., *Observing British and European Mammals* (London: British Museum (Natural History), 1989).

Bright, P., Morris, P. & Mitchell-Jones, A., *The Dormouse Conservation Handbook* (York: English Nature, 1997).

The British Deer Society (www.bds.org.uk).

British Geological Survey, *Chester (Solid geology map) E109* (Nottingham: British Geological Survey, 1965).

British Geological Survey, *Chester (Solid and drift geology map) E109* (Nottingham: British Geological Survey, 1965).

Bullion, S., *Key to British Land Mammals* (Shrewsbury: Field Studies Council, 1998).

Byerley, I., *Fauna of Liverpool* (Liverpool: published as an appendix to the Proceedings of the Liverpool Literary and Philosophical Society, 1854).

Carne, P., *Deer of Britain and Ireland: Their origins and distribution: Their history and distribution* (Shrewsbury: Swan Hill Press, 2000).

Carthy, R., *The Harvest Mouse in Cheshire 1999–2000 Survey* (Nantwich: Cheshire Wildlife Trust, 2001).

Chapman, J.A. & Flux, J.E.C. (eds.), *Rabbits, Hares and Pikas: Status Survey and Conservation Action Plan* (Gland, Switzerland: IUCN, 1991).

Cheshire region Biodiversity Partnership, Mammal Action Plans (available from www.cheshire-biodiversity.org.uk).

Churchfield, S & Brown, V.K., 'The trophic impact of small mammals in successional grasslands', *Biological Journal of the Linnean Society*, 31, no. 3 (1987), 273–90.

Cooper, V., *Marine mammals of the Mersey Estuary and the east of Liverpool Bay* (Liverpool: University of Liverpool, unpublished, 2002).

Corbet, G.B. & Harris, S. (eds.), *The Handbook of British Mammals*, third edition (Oxford: Blackwell Scientific Publications, 1991).

Coward, T.A. (ed.), *The Vertebrate Fauna of Cheshire and Liverpool Bay* (London: Witherby & Co., 1910).

The Deer Initiative (www.thedeerinitiative.co.uk).

Deer UK (www.deer-uk.com/Deer.htm, accessed 2005).

DEFRA, *Working with the grain of nature: a biodiversity strategy for England* (London: DEFRA, 2002).

—— *Working with the grain of nature–taking it forward: Volume 1. Full report on progress under the England Biodiversity Strategy 2002–2006* (London: DEFRA, 2006).

—— *Working with the grain of nature–taking it forward: Volume 2. Measuring progress on the England Biodiversity Strategy: 2006 assessment* (London: DEFRA, 2006).

Dobson, G.E., *Catalogue of the Chiroptera in the collection of the British Museum* (London: Taylor & Francis, 1878).

Dobson, J., *The Mammals of Essex* (Saffron Walden: Lopingna Books, 1999).

Earp, J.R & Taylor, B.J., *Geology of the Country around Chester and Winsford. Memoir of Geological Survey, Geological Sheet 109* (London: HMSO, 1986).

Entwistle, A.C. *et al.*, *Habitat management for bats* (Peterborough: Joint Nature Conservation Committee, 2001).

Evans, P.G.H., *Status review of cetaceans in British and Irish waters* (under contract to the Department of the Environment, 1992).

—— & Anderwald P., *Cetaceans in Liverpool Bay and Northern Irish Sea: an update for the period 2001–2005* (Oxford: Seawatch Foundation, 2005; available from www.seawatchfoundation.org.uk/publications.php?uid=11).

—— & Shepherd, B.S., *Cetaceans in Liverpool Bay and Northern Irish Sea* (Oxford: Seawatch Foundation, unpublished consultancy report to CMACS, 2001).

Gray, M., *Geodiversity: Valuing and Conserving Abiotic Nature* (Chichester: John Wiley & Sons, 2004).

Greenaway, F. & Hutson, A., *A field guide to British Bats* (Uxbridge: Bruce Coleman Books, 1990).

Greenwood, E.F. (ed.), *Ecology and Landscape Development: A History of the Mersey Basin* (Liverpool: Liverpool University Press, 1999).

Haines, B.A., Horton, A. *et al.*, *British Regional Geology: Central England* (London: HMSO, 1969).

Halton Biodiversity Steering Group, *Halton's Biodiversity Action Plan* (Widnes: Halton Borough Council, 2003).

Harris, S. & White, P., *The Red Fox* (London: The Mammal Society, 1994).

Harris, S. & Yalden, D. (eds.), *Mammals of the British Isles Handbook*, 6th edition (London: The Mammal Society, 2008).

Haussler, U., Nagel, A., Brown, M. & Arnold, A., 'External characters discriminating sibling species of european pipistrelles', *Myotis*, 37 (2000), 2–40.

Hossell, J.E., Briggs, B. & Hepburn, I.R., *Climate Change and UK Nature Conservation: A Review of the Impact of Climate Change on UK Species and Habitat Conservation Policy* (London: DETR, 2000).

Hutson, A.M., *Action Plan for the Conservation of Bats in the United Kingdom* (London: Bat Conservation Trust, 1993).

Introduced Species in the British Isles (www. introduced-species.co.uk).

Jones, G. & van Parijs, S.M., 'Bimodal echolocation in pipistrelle bats: are cryptic species present?', *Proc. Roy. Soc. London B*, 251 (1993), 119–25.

Jones, K. & Walsh, A., *A guide to British Bats* (Shrewsbury: Field Studies Council, 2001).

Leith Adams, A., *Monograph on the British Fossil Elephants* (London: Palaeontographical Society, 1877–81).

Lever, C. *The naturalized animals of the British Isles* (London: Hutchinson, 1977).

Macclesfield Biodiversity Audit (Macclesfield: Macclesfield Borough Council, 2006).

Macdonald, D. & Baker, S., *The State of Britain's Mammals 2005* (London: Mammals Trust UK/People's Trust for Endangered Species, 2005).

Macdonald, D. & Tattersall, F., *The State of Britain's Mammals 2004* (London: Mammals Trust UK/People's Trust for Endangered Species, 2004).

Macdonald, D.W. & Barrett P., *Collins Field Guide: Mammals of Britain and Europe* (London: Collins, 1993).

The Mammal Society, Species Fact Sheets.

McBride, A., *Rabbits and Hares* (Yatesbury: Whittet Books, 2003).

McDonald, R. & Harris, S., *Stoats and Weasels* (London: The Mammal Society, 2006).

Mitchell-Jones, A.J. & McLeish A.P. (eds.), *The Bat Workers' Manual*, third edition (Peterborough: Joint Nature Conservation Committee, 2004).

Morton, G.H., 'The Elephant in Cheshire', *Trans. Liverpool Biol. Soc.* (1898).

Neaverson, E., 'A Summary of the Records of Pleistocene and Post-Glacial Mammalia from North Wales and Merseyside', *Proc. Liverpool Geol. Soc.* (1942), 70–85.

Open University, *The Great Ice Age*, second edition (Milton Keynes: Open University, 2007).

Palin, N., *Alien Invaders—The Threat Posed to British Wildlife by Introduced Species* (2007).

Regional Biodiversity Steering Group for North West England, *A Biodiversity Audit of North West England, Volume 1* (1999)

Richardson, P., *Distribution atlas of bats in Britain and Ireland 1980–1999* (London: Bat Conservation Trust, 2000).

Rossbach, K.A. & Herzing, D.L., 'Inshore and offshore bottlenose dolphin (Tursiops truncatus) communities distinguished by association patterns near Grand Bahama Island, Bahamas', *Canadian Journal of Zoology*, 77 (1999), 581–92.

Russ, J., *The Bats of Britain and Ireland* (Bishop's Castle: Alana Books, 1999).

SCANS, *Cetaceans in the European Atlantic and North Sea. Report for 1994* (St Andrews: SCANS, 1994).

SCANS, *Surveys of the SCANS-II Project, July 2005* (St Andrews: SCANS, 2005; available from http://biology.st-and.ac.uk/scans2/documents/issue4_Sept05.pdf).

Sleeman, P., *Stoats and weasels, polecats and martens* (London: Whittet Books, 1989).

Stebbings, R.E. *Conservation of European Bats* (London: Christopher Helm, 1988).

Strachan, R., *Mammal Detective (British Natural History)* (Stowmarket: Whittet Books, 1995).

Strachan, R. & Moorhouse, T., *Water Vole Conservation Handbook* (Oxford: Oxford University Wildlife Conservation Research Unit, 2006).

Swift, S.M. *Long-eared Bats* (London: Poyser, 1998).

Tracking Mammals Partnership, *Tracking Mammals Partnership Update 2006* (Peterborough: Joint Nature Conservation Committee, 2006).

——, *Tracking Mammals Partnership Update 2007* (Peterborough: Joint Nature Conservation Committee, 2007).

UK Biodiversity Steering Group, *Biodiversity: The UK Steering Group Report. Volume 1: Meeting the Rio Challenge* (London: HMSO, 1995).

——, *Biodiversity: The UK Steering Group Report. Volume 2: Action Plans* (London: HMSO, 1995).

Vale Royal Borough Council, *Nature Conservation Audit* (Winsford: Vale Royal Borough Council, 2000).

Walsh, K., *Cheshire's Bat Records* (unpublished report, 1994).

Warrington Nature Conservation Forum, *Nature Matters! A Biodiversity Action Plan for Warrington* (Warrington: Warrington Nature Conservation Forum, 2005).

Waters, D. & Warren, R., *Bats* (London: The Mammal Society, 2003).

Wildlife Conservation Research Unit, University of Oxford, *Proposals for Future Monitoring of British Mammals An Overview. Issued jointly by the Department of the Environment, Transport and the Regions and the Joint Nature Conservation Committee* (London: DEFRA, 1999; available at www.defra.gov.uk/wildlife-countryside/mammals).

Woodroffe, G., *The Water Vole* (London: The Mammal Society, 2000).

Woods, M.J., *The Badger* (London: The Mammal Society, 1995).

Yalden, D.W., *The History of British Mammals* (London: Poyser, 2002).

Index